BioQuestions

and

the mechanical answer

Ψ

Didier Newman

Editor: CreateSpace

Bioquestions and the mechanical answer

© 2013 Didier Newman

English correction: Nuria Cohen

ISBN-13: **978-1484164259**
ISBN-10: **1484164253**

Bioquestions and the mechanical answer
Didier Newman

Chapters:

1

*It is only possible to travel if one asks on losing one's way;
otherwise, nobody leaves home, everybody remains within the answer.*

Ψ

DOUBT,
the origin and traditional biology

It is alleged that the origin, the beginning of life, is an unsolved brain game, a challenging puzzle without any established line segment; and so far, an always incomplete puzzle. A mystery that remains in the dark, still invisible to the third eye of science, which is vague, diffuse or useless to the definite clarification of this doubt which has been harboured. However, since questions always return tirelessly to the search, one can often conclude that science carries on working in this field of study despite not having accepted any definite hypothesis yet, which always must contain the truth, constantly subjugating some alternative reasoning, universally understood without ambiguities or contradictions; and hence, it has become part of the current theoretical body of biology or of any other branch of the traditional scientific knowledge related to these subjects. It is an adequate introductory example that, in spite of the fact that some fossilized living beings have been effectively identified and dated from about three thousand million years ago, there is not even a general consensus if it was on the planet Earth where the matter itself transformed until something that is now called life emerged and became present, as a castle would emerge at the top of a high mountain; indeed, a castle built with the stones of the same mountain, like sandcastles in the middle of an unspoiled sand beach.

This bewilderment is a coherent fact, the branches of the traditional knowledge are a set of tens, also deep and dilated; and as it is known, a lot of things are quickly summarized by the fact that a more powerful brain is needed to understand better the current working brain, or a set of more powerful brains are needed to better understand the set of current working brains, all of them part of the vital existence and its dilemmas. Likewise, it is also possible to have a limited vocabulary, or at least, all the necessary and sufficient words, or maybe it would be necessary to have more evolved vocal cords, tongue and mouth to articulate and solve this kind of problems properly. Certainly, at the same time, one requires an ear ready to listen without losing track, not getting constantly muddled with the simplest words and with the diverse and possible meanings. Anyway, are words a suitable tool for this business or are they like scissors used to stick two pieces of paper together? Are words, numbers, physical formulae, ideas, sophisticated mathematical concepts and the rest of knowledge objects, only an effort to humanize the universe, to anthropomorphize all things? Is this a very useless effort, a means without an end? A very noble effort or a vain attempt to dress what is in fact absolutely undressed? Even the simplest words which seem to be useful for everything and for the whole, like *this* and *universe*, are a ridiculous cover, an old-fashioned and precarious patch? Is this the fundamental issue, the weak and corruptible point of any kind of human reasoning? Do modern science and the logic of its language avoid this fact? Or in other words, do the universal, natural or physical laws and pure mathematics sustaining them avoid it as well? Is there any logic as pure as solitude itself? Otherwise, is silence the only real wisdom?

In any case, when there is a doubt everybody always gets talking, asking, even wondering about what seemed to be thoroughly chewed, swallowed habitually as a purée, baby food or a pill for those who are lazy; for instance, what is life exactly and what does being alive mean? Who or what is alive? What is sufficient and necessary to be alive? Is there a differential feature between the so-called animate and inanimate matter, an actual feature beyond an agreement between speakers and the catchy and limiting conceptual network of their language? Is life something that only exists for the observers who classify or is it there alone without them too? Indeed, is the observer part of life, also part of the things seen? Or as the well-known philosopher, philologist and Prussian psychologist said: *is the observer a hyperborean?* Meaning by that, is the observer an outsider, outiside of all that will become too, does he remain constant with every change? Is it something like a soul? By the way, what would be the soul of life, what does never change within it? Which is the fulcrum that holds the descriptions of its intimate nature? Is there anything that survives time in its moving structure?

These questions and many others, apparently trivial, can easily be left unsolved, indefinitely further and further in doubt or reaching supine ignorance; especially, when it is applied to a consistent scientific method, each step being more precise and inquisitorial, extremely rigorous with the statements and the solutions offered. Furthermore, if everything is mixed or treated with a relativistic perspective or if one includes the dense philosophical knowledge, the psychological mess, the philological punctilious reviews and some other severe and suffocating trends or modes of hard reasoning and inquiry. Thus, in parallel to the doubts that hover like vultures over a dying hypothetical vital fact, logically, any elucidation on an origin and its possible causes

11

is also fiercely questioned. It is doubtful whether it is possible to state any principle or any cause whatsoever. And if so, by using the same valid terms used until now and shunning the relentless danger of idealization, of diffusing towards the abstract or the absurd, or indefinitely appealing to a prior cause without causing a dramatic cut, as deep as the cut of an eye or an open wound – the one that would clearly allow seeing the outside or feeling the pain that demands an immediate cure from happy-go-lucky consciousness. That is, a cut as radical as the ultimate end of everything, at least of everything that is susceptible to terminate, to cease, to stop; that is to say, of everything that is mortal. However, what is mortal and what is immortal? Is there anything mortal or immortal under the sun? Is there something mortal that will not die, always fighting the watching death? Why, when, where and how, was the word immortal invented and did it have a future?

In order to get a little lighter and float on the rough sea of the next questions, some solid points can be established here as short nails hammered with only one blow into the wall of traditional culture, which still allows us to pose some inquiries on solid ground, at least in the past ground of current changing science. So, in the first place, a starting point of support could be the cell; identified in the academic tradition, in the world of the observers of the same things and with the same eyes and glasses, as the basic or fundamental unit of life, of living matter, either individually or in the multicellular aggregate form. Thus, the basic etymology of the word *cell* leads us – a few steps of a philologist backwards and before disappearing into the sounds of the animal voices of the Upper Paleocene primates – to the Latin voice, which makes it clear that prior to its current form and meaning, this word was used to describe a sort of small room, a vacuum achieved by the construction or the use of walls, which would allow obtaining a certain space with its own characteristics, for any particular inner uses, such as that of a hiding place, resting or drawing room, prison, fortification, refuge, casket, coffin, etcetera.

And now, without losing the analogical correspondence at all, in spite of having purged and added a great deal of new knowledge to the previous, the loanword is still useful to describe an essential part of the current biological theory, to describe this usually microscopic corporeal entity. This structured space that is separated from an outside world; at least, to a sufficient degree to identify a specific shape within the world at large. But, is not everything a particular form within the world at large? Doesn't a common stone above the ground have its own particular form? Is the total separation of two spaces within the same universe possible? If so, where is the universe confined? Is it just confined to one of the parts or is it to both but split in half? Hence, is any kind of separation always unreal? Is it only a convention that the universe itself inherently blurs penetrating within all things, happening all within? Even so, what is needed in order to separate successfully? Is some kind of vacuum necessary to separate? Is this the vacuum defined as the ground state of minimum energy? The vacuum of nothingness instead? Is the vacuum that separates things the vacuum of the form? Is *the form the vacuum* as is mentioned in some Zen and similar texts? So, is it always necessary to cover with a transparent veil of absence to perceive the figure, the pattern or the form without confusion, to perceive something particular and separated from the landscape that contains, nourishes and surrounds it? To perceive something emerged from the universe itself but still in the universe? Is this veil a little bit of vanity reflected in the sunlight? A little bit of shiny dignity perhaps? A little bit of conquered and self-assimilated nothingness?

Despite its fragility and delicacy, this property or feature, the clear separation of spaces, can be exposed here as one of the central pillars of what has been traditionally defined as biological life. And first, it is noticeable in the orderly autonomy of a cell, which manifests a shape or a pattern in a medium; a medium which becomes, simultaneously and due to the use of lipid membranes – equally when skin, feathers, walls or any another frontier or boundary is used depending on the level, layer, fold or scale pointed out beyond the lonely cell – also a surrounding or an environment. So then, by explaining a daring analogy, nesting it in the most familiar psychological area, the cell can be said to be the apparent manifestation of a tiny or large *I*, a tiny or a large *self* or *ego*. However, does the stone have any I, self or ego too? If so, what is its border or its contour, what is its personality? Does it have a very boring one? Or was the cell the first object that became a subject, or the first subject that become an object? Was it, merely, the first observer, the first to suffer from vacuum, the first powerful wrestler against an outside hostile world, against an amorphous and wild universe?

After all, the cell seems an active self, a busy room that is not totally isolated, it has doors and windows, empty spaces that are opened and closed, inward and outward, with some collaboration or by force. So, it has empty gaps that are useful, where some exchange occurs, a transport of matter and energy in both directions of the flow, inside out and outside in. Consequently, the cell also seems to run or to work on something; for example, homeostatic processes are taking place. And hence, selfish processes, at least for its self-sustainment, in order to move towards an ideal equilibrium, towards a certain form of health, as if it were one exhausted athlete who is resting, toning and recovering himself after running more than one marathon; to sum up, some sort of metabolism is carried out while time flies.

Then, the cell is usually presented as a space-form with content related to the outside, but unlike anything that is not a cell or a group of cells – such as a dissolved stone when it is embraced by the river of magma that runs down the side of a volcano – and using another psychological analogy to get out of the lava flow, the cell expresses some *desire* or *will*, despite being simply the will to shy away from the will of everything that surrounds it; or at least, from the appearance of everything that is around it. So, is the cell shying away from everything that is not itself? Doesn't it love fashion or uniforms? Does it shun old-fashioned coats of lava? But, is its own will, private and personal stubbornness? Or on the contrary, does it follow orders like a military piece, like a private? Is it an alienated entity? Is it a volunteer or a conscripted soldier without any choice other than war and death?

Certainly, parallel to this virtual will, the cell never stops to abide some kind of submission either. What is more, in the case of a relatively healthy cell, the degree of compliance is more evident when it is grouped with other cells to constitute multicellular living beings, thus becoming part of a gear assembly. Moreover, no matter what the final configuration of the living being is, unicellular or multicellular, healthy or ill, in any case, and without any known exception, there is always a present submission to the laws of nature: physical, chemical, and so on. Indeed, the law or laws, which cannot be proved, so far, that anything may escape; for instance, beyond theoretical probabilities, has an apple already been seen jumping upwards, without an effort from the ground, until it becomes part of the branch again? Does the cell or the cell grouping escape from natural laws minimally? Is it exactly what they want to do? Is escaping its innermost will, its primary driving desire? Is scaping the desire of life, escape from inanimate stuff? Besides, is there any biological natural law? Would this law be a basic part of the Theory of Everything or only a complex and folded reflection? Could the living matter escape from the law which it manifests as an actual example, could it repeal it? Is a stone a submissive thing too? Is it always quiet and comfortable, dissolving in the magma as if it were a masochist on holidays at a resort?

Leaving the cell aside for a moment, another solid conceptual anchor point, more modern in the history of science, is the *gene*. This time, the etymology leads us to Greek and Latin too, but one can get from both languages, with the root of the word almost intact, to the Indo-European vocabulary. Then, scattered across the globe and time, the basic root *gen* is known to have had a lot of multiple related uses or meanings, very similar but not identical, such as giving birth, produce, seed, birth, race, caste, family, lineage or newborn; finally, the linguistic loan was successfully introduced into the biologic cultural field after being adopted by the members of the domineering scientific world who were studying the biological evolution shortly after Darwin. In the meantime, as it is known, the same root is still, and long before the mentioned adoption, in countless commonly used words of many living languages, such as in the current English words: origin, genitals, generation, genocide, and so on. And therefore, any of the speakers of these languages may have a very vague and general idea of what is being treated in the specific jargon; even though they do not know anything or only very little about the existence of the actual gene, which is mainly in the cell nucleus and inside mitochondria.

Then, in order to start giving a unfledged explanation about that, the gene has been identified and traditionally defined as the basic unit of heredity of living organisms; although, it can be specified as the basic unit of legacy or bequest too, at least when it is not implicit in the first term or when there is no fear or risk of incurring in unwanted liabilities. In any case, genes are inherited and bequeathed directly, mostly representing the baton between generations, the thread of life. Thus, singularly and ideally, the gene is a biological piece that endures and changes over time; and therefore, it is also the very essence of the evolving flow, legally responsible for the fruits of the tree of life. Besides, it is a biological piece that is not transmitted vertically from parent to offspring alone, but also horizontally by mobilomas – genetic entities able to move around the genomes – either moving within a single genome by a transposon or between the genome of brothers or cousins in direct contact; for instance, in bacterial conjugation by using plasmids.

Nevertheless, without delving into the vast current genetic science, more than unapproachable for the individual – who with his hominid arms would have to embrace a huge banyan forest – the gene, at least its simplistic conceptual abstraction, leads to developing the inheritance and transmission concepts more accurately. So, it is not a mere baton like the one used in an athletic relay race, where a runner gives it to another runner who takes it with him and goes on to pass it on again, losing it on the ground or until it reaches or crosses the finishing line. It is not just an heirloom or a debt item, a large or small quantity that is undeniably and relentlessly transmitted if one wants to continue with the race. The gene, or more precisely, the specific variant of a gene – the allele – is a sort of magic packet, a present with something inside that does something; namely, the use of its *content* immediately derives in the constitution of the runner's feet and legs; because, it is the carrier of building instructions, the instructions for RNA and protein sequencing; that is, the gene products that aim to rule, like good children of Roman emperors or strict resentful teachers, the structural and functional expression, the constitution and development of any of the participants of the race, perhaps with their actions and reactions included, both the reaction to victory and to defeat. Obviously, this is what is transmitted and inherited.

Thus, just like pieces of biological information, the gene inserts a dagger straight into the heart of theoretical life. Besides, the information – its storage and transmission, its structure and its use – becomes one of the fundamental pillars of modern life sciences. Perhaps, only because it simplifies the understanding, biological science is almost completely stripped of all other dresses and it is dressed again from head to foot with genetic lingerie, following a better twinned fashion with the contemporary world, with the age of information and media, computerization and technological derivatives such as the computer and the whole world orbiting around it. In fact, one might even think that everything which is of some value and is worthy of communicating is nothing more than a means to inform, some automatic activation or expression of information, which is treated as an end in itself.

Nonetheless, is this the way genes act as well? Furthermore, is the entire universe nothing more than expressed information, with or without any meaning, occasionally or always meaningless? On the other hand, is the universe a single meaning, leaping and transforming between synonyms? Is the whole universe perhaps information without it being expressed, still waiting for a careful observer to formulate words with the *word search*? But, isn't information all that can be used as a means to inform, any duality or any multiplicity that is susceptible to be selected? Is duality the minimum condition necessary to inform? Is informing simply pointing at one of two or more different things? So, is information simply the difference? Is expressing or communicating something as simple as pointing the index finger at black or white, zero or one, life or death? What information is provided and what is received when a finger points at or chooses between two or more identical objects? What is the information given by the mimetic animal? Does it say: "I am the other"? Is it possible to receive information without acknowledging what it is, as the proverb states *while one is looking at the finger the other is pointing at the moon*? What is information without communication, where does the word that is not heard end up? Does it end up in the mouth? In fact, is it possible to distinguish something equal from what one is? Is it possible for a human being to tell apart another human being? Is it possible to distinguish simultaneously the hand from the arm, the finger from the hand, and the nail from the finger? Is information something physical? Or is it rather something metaphysical? Is ignorance a metaphysical state? Are metaphysical aged problems a part of modern sciences? Are they now in the field of current psychology and neurobiology?

Once the gene has been introduced through some questionable, precarious and express literature, the load-bearing walls that constituted the current traditional biology can be cemented as well, building them on the site of the gene. Hence, it seems appropriate to expose superficially, quickly and compressed, also with the inevitable shortcomings and gaps, some other basic concepts such as *mutation* and *evolution*. Both words, part of the conceptual context which describes the molecular informative changing world. That is, based on the navigation adventures of genes over time, constantly storing and transmitting instructive information without the need of a neuron brain or a mouth with lips. So, in this way, by telling this story as if it were a remembrance, it can be said that the genetic information of an ideal gene of an earlier space-time was modified because of a mutation. Hence, a chemical change that directly affected the structure and basic instruction of that gene – other types of genetic changes are possible too, such as epigenetic changes or genetic drift, although there is no need to comment them in detail now. And, from this crucial event, some necessary conditions were established and the biological evolution began its journey through the desert. Certainly, this and other mutations were transmitted successfully through asexual reproduction, recombination, and so on. Therefore, changes started to take place in the genetic information stored and developed in new generations; in turn, prepared to suffer and to do the same thing repeatedly with the exception of a few.

But, what is known about mutations at will? Are genes like a punching bag, do they mutate after a low blow? Is the biological set only a flow of changing information? Is this a *Samsara* of information? Are past, present and new living beings completely reducible to information? Is this information totally computable? How is new information added to the genome, how do the set of genes grow and multiply? Perhaps, is an elementary gene copied several times, every copy mutates quite a few times more and finally the outcome recombines as if it were an orgy? Did evolution start with the recombination of one original copy and one mutated form of the same previous gene? Something like a biblical organism recombining itself fondly with one of its ribs, loving itself out of paradise of immortality, out of exact and indefinite reproduction? A sort of transcendental masturbation? Otherwise, did it start with two mutated copies of the same gene, two ribs or two complete new bodies exchanging points of view about passion and romanticism? Whatever happened, did evolution begin with life about four billion years ago in an idyllic or horrific place? Or is evolution a universal law that captures the entire universe and in consequence also inevitably captures life, like a dry leaf that the wind blows away?

Even today, evolution is for some people a flight forward, constantly jumping into a dark abyss full of dangerous thorns; as a result, the conservation of working machinery and its associated information is continually and drastically sought, since the hostile environment survives tirelessly and is on the lookout, no matter how wide, high and safe the lipid membrane has grown. Indeed, one wants to avoid at all costs the cold choice of death, that is stubbornly the same in a diverse and changing world that is heading towards the blows of its scythe; however, after its precise and unavoidable cuts, a sort of selection becomes apparent, a natural or artificial selection, being a selection in the sense of avoiding or making it possible to procreate new offspring or to repeat exactly the same in a new world. So, among other things, in this way *diversity* is limited and channelled, some information is deleted from the maps as if it were written in pencil and the rest is located in specific scenarios, like dolphins in the seas or aquariums and lions in the jungles or zoos.

Thereby, stubborn like an old programmed machine, death always favours those among the weakest and the most reckless, among the altruistic and the sick, among misfit genes or alleles; hits the target of obsolete originals and failed mutations of a specific time or space. But, that said, is death a programmed machine or is it rather a free wild will? Is it a sort of machine programmed for savagery? Or is it a wild code that is programming all kinds of machines, all kinds of hardware? What do those genes that walk without an umbrella think about it when giant meteors fall on Earth? Or what about those tormented alleles electrocuted by lightning? What about those that carelessly allow them to have a bomb dropped from a military plane? Is lightning more natural than a bomb that free falls from a military plane? Is it possible to have a natural selection yet? Has a human being or a rebel chimp ever changed the world intertwining all subsequent events? If so, is everything that occurs from now on to the future inevitably artificial? Has the world ever changed by means of an absurd event? Then, is current life in the era of an absurd evolution? How many types of selections and evolutions are in competition now? What will animals, plants and fungi, bacteria and archaea be like once they have survived the postmodern artificial selection era? Will they also be artificial? How will a wild lily survive the ideological selection, what will its flowers be like once the ideologies have ended? Will their petals look like flags? What other sort of adaptation could be useful for the lily? Equally, does economic selection affect bacteria at all? Will those that work frantically and in an altruistic manner be the only ones that survive? Along these lines, has the weakest, poorest, the most misfit, the illest and the unhappiest of all living beings in history ever reproduced? Otherwise, has the strongest, richest, the most spoilt, luckiest and healthiest son of Earth always reproduced? If so, who inherited his or her fortune? How many heirs receive a portion, was it a tiny or a big portion? Could the reproduction of this kind of Superman fail?

Besides, in this context of mutations and mutants, obsolescence and fitness, success and failure, fortune and poverty, good and bad luck, what was the first message, the first instruction, the minimum information effectively stored and transmitted? What was the first step towards the building of the house? Express the will to build it? Demolish the previous old house? Lay out the foundations and the braces, a specific protein? Look for an architect to draw the blueprints and to work out the structures? Bribe the Council to make it all illegal but with the corresponding authorization? Hire a mason, a ribosome? Has the hypothetical original information been totally preserved amid the mutations of added information that has taken place over time? Is the essential information of a primary basic cell kept original? If so, does the transformation of the originals occur without observing copyrights? Can one say that biological information is also some kind of knowledge? Is it some kind of wisdom too, maybe a little living truth? Is it rather an incomprehensible foreign vocabulary, an anonymous mathematical or chemical hubbub? Or is it tactful and sweet words like the melody of the mother tongue? But, is communication the use of genetic information? Is a nose, a heart, a horn, a femur, a finger or an eye, physically expressed words? Something like pulling a face to the neighbour or giving him the finger?

Anyway, the genetic material constitutes the key for the understanding of the present biological world; accordingly, also for considering any speculative composition about the origin of life and the future of it. For this reason, the study of biological information, its possibilities and limitations, is carried out fervently, more than ever and by using a huge amount of resources; whether this information is part of an old tradition, such as the genes that usually participate to make birds with wings, or conversely, new information, such as recent mutations or any other kind of *change carriers*; also, whether it is the test or the experiment of an inventor or the artwork of an artist – conscious or unconscious, real or imaginary inventors or artists, experiments or artworks. But, is life an experiment, an experiment made by a nutty chemist? Is life itself art, the artwork of a misunderstood visionary genius? Are genes some sort of artists? Is there art in the body or in the beauty of the flesh? Is it rather an experiment or an artwork without a method, like a child casting a spell or painting the wall with a felt tip pen? Was the beginning of life a radical mutation of the universe, a complete change in the displayed language? Does the universe have some tradition or is it always new, with no debt at all? Is there any information beyond life, genes, technology or written and oral traditions? Is there any information in outer space, whether there is life or not, whether its parts have genes or not? Is there any information in stones? Were stones formed through information processing? Does the information contained in a stone make any sense or is it just a word that ends in its shape, dying in the lips or in the tip of the tongue? Does the voice of the gene end its journey in the form, in the protein, in the phenotype? Is a face something more than a carved stone? Or is it a stone with eyes watching and carving the surrounding air instead? Do feelings, behaviours, reasons, dreams and aspirations have form? Could this form touch or be touched?

Precisely, the full information stored in genes makes up the genome, the specific messages of a unique code. This code is generically called *genetic code* and it is currently used in all known living beings. Thus, it can be said that the various discrete manifestations of this code, once they have been selected, edited and published, allow geneticists to differentiate forms of life from each other and from their ancestors and consequently, they make it possible to trace filiation and evolution too. However, one must also take into account that genome cloning can also take place; meaning by that, the exact copy or the reproduction of it, this sort of immortality seed. In this case, there is nothing to differentiate at all and it is necessary to wait patiently for autumn, up to the first grain of sand that falls from the tree of time; and only then, while the present moves and diverges by itself, the path of clones moves and diverges within. In that regard, is every living being one of these diverged paths too? Is it perhaps the pedestrian travelling along them? Is every living being one diverged creative space-time? Is it one personal and walking event? Is every current living being a new path and its walker, both diverged from the path where a bored clone went for an endless walk? And now wanderer, is there no path? Is the path made by mutating? And on looking back, does one see the path that will never be cloned again? In other words, wasn't that said by the poet Machado too? What do poetry and science have in common?

As cited before in this short path made so far, now made of footsteps alone, the baton of the race, the gene, at least when it is up and running, participates in the gradual formation of the runners themselves. From the innermost nuclei, it takes part in the arrangement of the matter that already exists; as a result, it participates in the description of another basic life trait effectively, which can never be omitted, and which was already implicit when the separation trait and the relationship of the cell with the outside world has already been mentioned. This trait is the domain trait. The ability to control, to have power over the environment, to transform it in a certain way; often reduced with the help of thermodynamic science to the basic self-directed energetic process in which the entropy of the own system is diminished by means of increasing the surrounding one, which can also be regarded as feeding oneself to survive at the expense of others, chewing lettuces, ripping flesh or parasitizing someone else.

Therefore, when power or dominion is sought in the interior of living beings, the genome seems to play a main role, since genes carry instructions, interrelated mechanical orders or commands that eventually form an organism, a body; in turn, also with a domineering tendency. Certainly, something that naturally manipulates an environment which exhibits natural resistance at all times and places, scales, layers, levels, folds, and so on. However, is dominance a necessary condition to achieve the mentioned separation, the separation between the cell and its surroundings? Are powerful hands essential tools to erect a stone wall or a lipid membrane? Or is dominance a trait that is just derived from separation itself? Only after separating the hand from the stone, can one handle stones to erect a wall? Is this perhaps a single process, with no switching parties or alternation? Does it hide a reference paradox or is it just that cause and effect, the former and the latter, are sadly confused? Whose wall is this? Whose hands are these? Is it a shared dividing wall, a party wall? Are there dividing hands involved too? Otherwise, are there shared hands, hands building the wall from both sides?

And, meanwhile a powerful dominating life is presented, arranging or subtly shaping matter, and this being an essential feature shared by every living being, it should also be considered, what is needed to dominate and why is there this need to dominate? Is there any need at all? Leaving aside the subjective hunger, is there any need to go hungry, to need something? Is there any need to live or any need for life itself? If so, what was or is the lack that life or living beings replace? Is necessity the first life characteristic, is it at the top of the hierarchy among vital traits? Do stones need something? Must one need to exist or must one exist to need? Directly, is existence a state of necessity? Is life a perpetual necessity? Is life necessity and pure and absolute dependence, a prototypic parasite? But, is life the parasite or is it the host of a parasite that sucks like an insatiable leech? Furthermore, is life an absolute need with no possible satisfaction or is it the source of absolute self-satisfaction, power itself?

These key characteristics of life can be related to thermodynamics science too, which again helps reasoning with its second, more famous and as we can see, milked law. A law stated here, diffusely and far removed from its specific field of knowledge, as the natural increase of the entropy of the universe or the universe positive entropy variation; that is, the natural tendency of godforsaken systems evolve towards the lowest energy levels or become disorderly until they finally sprawl in equilibrium, like utterly hopeless and defeated losers. Or, as it can be experienced daily and clearly, how easy it is to make ten thousand glass pieces of all shapes and sizes from a beautiful glass figure, only by opening a hand and letting the glass figure drop, however how complicated and laborious it is, from the ten thousand pieces of glass, and only with the same hand used before, making the glass figure again. What is more, rebuilding the same identical glass figure without adding glue to the equation, without scars of any kind.

Hence, as a good friend of neglect and destruction, this slightly expounded law implies that any proper or autonomous trend or will, such as the case of the cell, has to swim against the tide and continuously fight, struggle, since it takes its own time to make slender figures of glass from many rough and scattered pieces. Perhaps, many more pieces than one can see at first sight or using the electron microscope; for this reason, it seems that there is a price in order to dominate, to arrange or to configure, to form a structure that does not stand alone, with a certain personality, a structure that is always calling like the newborn bird from the nest, with its mouth wide open, always in need. Then, life looks like a parasite of necessity and necessity looks like a host of life; and obviously, if this is so, it was a pool of necessity before the living journey started. So, from this point of view, the beginning, the biological origin had to be something like a gift or help, an effort or a sacrifice, the payment of a passage, of a levy or something similar. But, was there really an origin, an initial face-to-face with the Second Law of Thermodynamics, a confrontation that is still going on? Is the initial expression or manifestation of this conflict the origin of life? Was the passage or the levy paid by means of a loan, paid through the remuneration for some work, through the spoils of a theft? Who or what made this effort or sacrifice, was it the universal banking system or a particular client who lacked an organized structure to get dressed? Did someone sell his soul to the devil in exchange of a fragile body? Was an insane deal signed, having a green soul for today in exchange of giving it back more mature tomorrow?

And after a while, life began to evolve and is evolving into increasingly complex forms; as a result, there is a deeper necessity which takes more and more from outside, from the resistant and undomesticated surroundings. Then, a much greater control and sharper environmental submission is also needed. However, for both small and large loans, does the wild surrounding look like a commercial bank? Is it a cruel environment that lends its resources whereas it is waiting armed for the yield of an interest? Or is it a financial bank, a highly speculative one instead? Was life a banking initiative, looking for a debtor to do business? Or was life the initiative of some inner desire, a sort of image to be obtained from the lack of the image itself? Had the cell imagined itself before becoming manifest? Was it like one who wishes to buy a house and then gets endorsements, asks for a credit and finally purchases it? Where could an imaginative cell have seen a cell before, knowing what it was like? Did the cell randomly buy a costume in a body store? In any case, what is life seeking, what is the desire beyond itself and its own appearance? Which is the desire of evolution, what sort of meaning do living beings attach to it? Can they be objective about these matters? What is the most essential need and what is the most superficial fad? Is energy this essential need, just energy disguising itself as money or power that allows things to be done? Are bank savings also energy, locked away while keeping an eye on inflation? Will life end by consuming all the energy of the whole universe in an absolute ecstasy in obesity, doing away with the bank pocket? In what manner will living beings be fed when there is no more environment to consume at the peak of evolution towards infinite complexity? Then, will life feed on itself, on its own fat? Is it not already feeding on itself? Is not being alive always feeding on oneself? In turn, with no food to eat, is life a stake, a bloody mess, a bluff towards bankruptcy, a bet without good cards that will fall off a cliff when least expected? Is hunger betting life on an *all or nothing* game, a continuous *all in* bet? Is life an irrational adventure that started from irrationality? Can anything rational arise from irrationality? Is rationality something more than pointing out irrationality?

At this point, a bit better placed among lost objects, many more questions can be asked to unleash or to leash controversies, also to awaken the spirit of inquiry and perhaps to allow some knowledge to sink in. Or rather, to allow the reader to recognize, if it has not been done yet with the few previous questions made until now, the ignorance and the imprudence of the narrator or its sources, those essential facts that are not known and those that are being totally misinterpreted. Anyway, more questions of all kinds can be put forward here; relatively interesting, contradictory, ambiguous, meaningless, repeated, twisted, deep, innocent, ironic, funny, necessary, foolish, and so on. Even as if they landed from the beyond to blur and mix the narration inside subjectivity and the firmer and former traditional culture, also within more artistic and less strict, axiomatic or dogmatic views. However, despite these questions being introduced without identification and trying to talk about everything, a lot of them dealing with biology and other scientific similar subjects too, the same subjects taught in most educational centres around the world, both according to the current era and in an old-fashioned way, with more or less exactitude, forgetting or remembering the whole essential fact of living while studying life. In any case, one questions without any conscious restraint or eluding some previously excluded area as if it were an infectious disease, only hampered by the inertia of the past, the unknown limitations or the known limitations for the time being impossible to defeat and overcome.

For instance, as more than one may have already pondered, is the beginning of the *Gospel of St. John*: "the word, the verb, the logos became flesh", a literary narration very similar to the working mechanism of gene expression? Maybe, do the Gospels need to specify, somewhere, flesh as protein? In this same context, is the concept of incarnation appropriate, although it may not be wisdom which is embodied and shared with the rest of incarnates? But, can one assert so lightly that information, specifically the instructive informative message, which *in-form*, which brings form with or without an emitter or speaker, was and is prior to protein, to flesh? Is it also prior to any known matter or energy? Is there another type of universal code in this case? As Charles Bukowski said: *when spirit wanes form appears*? Is it so? Is spirit the information stored in genes? Is form the hangover after a spiritual night? Does genetic information feel degraded when it is embodied in a protein, as if a human being was embodied in a worm? Or quite the opposite, is it as if a worm were put in a human dress? Aren't human beings or worms nothing more than their bodies, nothing more than dresses made of atoms, carbon and so on? But, is there an informer or an initial source of this information? Is an abstract code without contingency with its effective constitution a necessary cause of a genome? Something like platonic grammar that can be used to instruct, to make hands and feet, but only becoming manifest when it is formulated with inky molecules? Are discrete works of this code incarnate or reincarnate, are they individuated and stored in diverse supporting entities such as RNA or DNA molecules? Originally, were these biomolecules just a product of noise, music coming out of the piano of the universe by randomly touching and touching its keys as if they were pressed by a monkey without music studies? In any case, once present in a physical form and attached to evolution, is the gene and not the cell the most adequate theoretical and ideal basic unit of life? Is a virus or a plasmid life too? Was the gene, a specific gene, the initial form of life? Was it the entity, the golem, the dust that inhaled the breath of life? Was the breath of life blown by something alien and transcendent, something already alive? Was it air of a spiritual or material code, free laws in heaven or slave laws in hell being executed on Earth? Or was the gene or the cell, a single emergence of a single random plot? A random outcome of deterministic laws that launched life as a mechanical expression, like a sleepwalker travelling at night? Is *random* a valid concept for a serious deterministic perspective? But, are there words that can express life and its origin and simultaneously be heard and understood? Can a seed or a living tree be heard and understood? Is there something to understand or comprehend at all, are energetic words expected from the deaf and the mute? Is energy the daily bread and not only on bread do human beings live? Is the Christian cross also a useful analogy to describe the basic atom of life?

So, is this carbon atom representation: $-\overset{|}{\underset{|}{C}}-$ the carbon on the cross? At what level of knowledge or confusion can we find theobiology science now? Is the science of its theoretical body doing art or technology? Can theobiology answer without dogmas any of these questions?

2

Breath and voice remain, transforming the present tirelessly.

Ψ

THE PAST,

the genotype and the common ancestor

When the cell is defined as the basic structural and functional unit of life, the same definition forces the exclusion of the solitary gene, or an isolated set of interdependent genes, from the private property of typical life, which would be life with a face and behind a face. Thus, in the formal approved dictionary of this life, genes are simply parts of a cellular mechanism. Accordingly, after passing through this tight filter or through such a selective prism, some voices often exclude entities with genes and not much more in their backpacks, such as viruses – marginalized pieces of life, outcasts or pariahs looking for insatiable revenge and opportunities – or plasmids, which like viruses do not exactly meet all the characteristics of the cell, which represents the complete macromolecular machinery, the minimum living being, the simplest face.

For a more lenient and permissive contemporary view, which has already seen and heard too many things to be surprised, these seem narrow points of view behind fogged glasses; in turn, all immersed in a very dark context, where one can sell a spider without a leg for the price of a stone with a spider leg taped on. On the whole, it is known that the invariable persistence of the contours of traditional concepts and their semantic content prevents or delays the timing of biology with new prevailing scientific paradigms. Consequently, for example, a theoretical and biological language unconnected to relativism and quantum mechanics seem to prevail, also ignoring the string theory that aims to bind the bossy worlds, and so on. Equally, unconnected to modern mathematics that underlies and sustains all the physical models, increasingly abstract mathematics, increasingly detached from everyday experience, with many formal and intricate artefacts supporting on each other.

Certainly, not only biology but science itself while it makes a breakthrough, needs another language, another grammar at the end of the day; in short, another world at its feet, a new complete world of concepts that in order to be conveyed, written and listened to slowly and carefully, has to become masticable as well. However, will teeth become loose along this tough road, will they become vestigial appendages? How can one bite, swallow and digest something abstract or undefined? How is something truly abstract built? Is it still something? Anyway, when embracing a new conceptual and scientific language, does another kind of life appear? Will another way of life pop up? Another kind of living being perhaps? A new observer and a new observed universe? A completely new universe? Has anything changed in this regard since fire was controlled about five hundred thousand years ago? And from the time it is known that a free tiny piece of the Sun, captured by a green plant in its growth using a photosynthesis process which includes quantum entanglement, is later released by warming and emitting tailor-made heat and visible light in a sweet home? Did biological life change at all when Albert Einstein proportionally equalled mass and energy in a physical equation? Did he also match bread and body, genotype and phenotype, will and pleasure, information and form, compassion and suffering? So, is will equal to pleasure times the speed of light squared? Is phenotype equal to genotype times the environment square? What is the strength and stability of classical and physical concepts such as *energy* or *mass*? Is it based on the invariability of the speed of light for all observers and speakers? In the same way, are the *equal* mathematical sign and its physical concept still meaningful? Why this constant need to equal? Is it based on previous prevailing confusion, perhaps on injustice or inequity? Is equality needed for clarity, in order to converge diversity or multiplicity into only one thing? Is confusion, injustice, disorder or inequity a natural state, more natural than to be born naked? Is confusion living within unceasing multiplicity, within this and that? Is the current scientific language no more than a weak and ephemeral agreement, a threatened dying man living on borrowed time near a precipice? Therefore, is will, compassion or bread also energy? Is any kind of diversity or multiplicity just a scam revealed by science? What do we know at all, and what are the changes between the past and the future with each new step of additional knowledge?

Halfway through that past and any sort of future, biology has also been strongly shaken by the new knowledge that it favoured before. Especially, when the vast and radical importance of genes in the biological whole set was understood. The leading role of the so-called information, its selection and management; it used to form, keep up and to transform all living organisms and vital processes. Hence, with genetics, the biological gears were lubricated to travel towards new biological paradigms, more coherent and adjustable to other modern scientific trends; indeed, a firm and solid node for transversality, its own anchor to cast the aid of foreign knowledge that is offered, bought or begged for the new understanding of life sciences. Thereby, genetics becomes a tool to check the way how biology can be understood under the big umbrella of information theories and how living beings fit with other scientific fields closely related to information itself. So now, what underlying new implications does biology conceal? But again, on consolidating a new biological paradigm, does it lead to a new path for living beings? Can current biological paradigms be extrapolated to the dawn of life? Is the origin doomed to change as the world makes progress? Does avant-garde science change history with the new understanding of the world that emerges from it? Is nothing as was thought because nothing real can be faithful to limited thoughts? Is it possible to establish any paradigm at all in a changing world, in a living world that does not allow anything dead to lie on it peacefully? Is change the only law that does not let one down or does change also changes its appearance, its name, its face?

Nevertheless, even under the stroke of the lash of relativity and of the absolute uncertainty of any unfocused approach to the facts – maybe with a crossed gaze behind the eyes looking inside the skull – information allows us right now, just to keep the narrative going on, to expose a minimum consensus between the two traditional units of biology: the *gene* and the *cell*. This consensus is a precarious conceptual idealization that initially describes the vital basic entity as a dual object; meaning by that, an object where these basic traditional concepts confront each other, face to face and at the same level. For the lovers of compound words without reasonable prejudices, maybe a new useful concept that one can name, in the zenith of ironic originality, as a *gell*. So, a new biological object is presented and it can be understood as a duality; or rather, if one goes deeply into whatever dual nature, a relationship between two parts. A relationship which, in this particular case, is simplified or summarized for some kind of logical agreement, as the genotype and the phenotype relationship, making the usual

34

meanings quite flexible to the point of breaking them into new concepts, which will arise from the ulterior use of words. Therefore, the *gell* would be this relationship, simultaneously a genotype and a phenotype. Although, we should not forget either, that if it were not an isolated relationship, it would be a relationship always sabotaged and tailored by a third party, by the influential and demanding environment, constantly bothering and drawing attention to itself. Obviously then, as it can be deduced, this harsh environment was excluded in advance from being implicit in the ideal and popular simplistic definition or equation of the phenotype: *phenotype equals to genotype plus environment*. So, taking into account the third actor playing in this live movie – apart from the ideal *gell* concept – this small and ideal mini-ecosystem can also be exposed, a minimum universe with life; a synthesized system that does not hobble for a while, like a three-legged table.

Along these lines, expressing some kind of solidarity with the anxious and annoying questions of the gutter press, is it necessary to specify any intimate details about this relationship? Is it the relationship of a loving couple and the enchanted house where they are kept in captivity? Are they a couple living locked in their own home? Is it a squatting couple? Is it a young married couple and an old priest who watchfully chaperones them everywhere? Or a married couple in a registry office and a law representative that also chases them with laws and bills? Are they a genotype and a phenotype kissing and holding hands in a romantic and idyllic garden? Or on the contrary, is it a couple living with a troublesome guest in their house, as if it were a pebble in a shoe? A couple and their parent, are the genotype and the phenotype two brothers? A dancing partner and the music that bewitches and leads them to shared loneliness? Perhaps, is it a *ménage a trois*, a love triangle, a trio without any hang-ups or obsessions? Maybe, is it a *duplex* or a chain with three handcuffs? Is it an unbalanced couple continually confronting everything, both of them against everything else? A couple at war against the environment, with no help and unshielded overcoming the constant bombing of the Second Law of Thermodynamics planes? Or maybe, is it a much calmer, older and post-war relationship? Perhaps, a three in balanced relationship, a system obeying the Zeroth Law of Thermodynamics? Besides, is this a pure and straightforward relationship, with no related parties and with no polar ends? Or is it a close, face to face and personal relationship?

Apparently, when studying the sequential process of gene expression, the complete chain of events of the *informative process* or *information processing* – for the time being: transcription, splicing, translation and finally post-translational modifications – it is observed that, by means of using a genotype and after negotiating downwards the contract terms with the rest of the universe, a phenotype is finally obtained, a gene product. Hence, leaving aside the strict surrounding for a moment, it can be highlighted that information is copied, interpreted and selected, combined and discarded, chewed and spat out, applied and sculpted, at will and with no hindrance. In other words, an ideal process of reforming and running information is carried out to achieve a desired goal. Briefly and weakly explained, as if oxygen were scarce and one was breathing one's last breath to reveal the name of the mysterious murderer before expiring: the phenotype of a genotype uses this genotype to form new genotypes and phenotypes, or newer ones. Namely, using the twisted logic of linear time and masculine filial analogy: the living son of a father, who lives in him, uses this father to make up more living sons.

But, who came first, the child or the parent, the egg or the chicken? Is the egg the son and every father was an egg before? Who is who in this ideal cause and effect relationship? Can there be paternity without children or a child without a parent? Can there be chronological time with no filiation between moments? Can there be chronological time if the filiation is only between brothers or sisters, will that be the eternal present? Are there also parents and children in the inanimate world, beyond life filiations? Is it possible to apply the terms *parent* and *child* when studying and interpreting the whole cosmology of the universe and its origins, or is it a heresy? Is it understandable that a bacterium and an archaea, a fungus and a plant, a lizard and a whale, all of them have their own parents, but the entire universe that contains and forms them, doesn't have any? Doesn't everything have to obey the same physical or natural laws? Isn't the universe as a whole under the same law? Are laws the mother of the universe? Are they perhaps its daughters? Is there any pure and unknown law beyond the projected shadow, changing over time, that science follows everywhere to trace its contour over the blank ground of knowledge? Anyway, is filiation the beginning of life? That is, is the beginning of one the beginning of the other?

Thus, addressing the issue as one who is drowning amid the thunderstorm and boards a sinking ship, to continue the adventure a little longer, a living being is understood here as a complementary and dual genotype-phenotype object, information and form, genome and expressed form – traits or features. It is simply another perspective. Nevertheless, also immersed in a mutiny on board, don't we need to define better who finally expresses this information, the entity that combines and transforms one thing into another, air into voice? Ancestral memory into genes, genes into faces watching the present? Is this entity real, does it exist? Doesn't it play a key role in this relationship? Is it a leading or supporting actor or is it an extra cut off from editing in post-production? Is it a wise perspective to identify this leading actor with the phenotype alone? This being reduced to specific cellular organelles which actively participate in the production of genetic products? For instance, this being reduced to ribosome, to the factory that assembles proteins? Isn't a ribosome made of proteins too? Is biological logic lost and blurred, wrinkled, twisted or folded throughout time and complexity, seemingly tangled in the present? Is the environment itself, the third actor in the whole relationship, this transforming entity? Is the environment the supreme manager of production machinery or is it a slave who works at the pace of lashes perpetrated by a cruel couple? A cruel couple asking for the impossible? Who or what is the artist who takes the genotypic clay and makes a phenotypic pot with handles? Is it time, dressing and undressing clay? Does time have artistic hands or is it only concerned about moving things towards the future as if it were a horse pushing a cart full of coal?

Despite the difficulties to find out what the original ingredients of a mixture with an original look are, like cola drinks, and despite the intricate order of causality where effects also become causes of other effects, as if it were a spinning wheel making fabric for a dress, this genotype-phenotype duality can be established, living and struggling, moving with some apparent alternation over time. In summary, a relationship of interdependent parts, worthless and irrelevant without each other, as is the case of the sender without the receiver, like a fairy tale without an imaginative reader; as they are the parts of the original symbol, that single object that was divided into two so that it was necessary that the formed parts were re-joined again, retrieving with that the value or the meaning that remained only within the whole object, in its unity, like a bank note torn in half and given to two different people. Or, although it was inconceivable for his real mother, as the little boy whom King Solomon hinted he would cut in half, to divide the two halves between the two possible candidates who were claiming him to be their son.

Incidentally, could the *gell* be used as a symbol, being the genotype and the phenotype their broken parts? Can anything be used as a symbol when one is the candidate and not the true mother? What is indivisible or conditions its existence to indivisibility? What is it that can never perform the function of a symbol? Do all divided things lack full meaning? Are the cells that divide themselves through meiosis also following this paradigm? Is a sexual body ever complete? Is the whole chastity of a gender complete? Are massive objects a whole unity according to the tyrannical will of gravity? Are masses parts of a jealous and prior whole that is pulling from the past? Or are they parts of a jealous and subsequent whole that is pulling from the future? Is this jealous being the present instead? Is there anything without a relationship, without participating in any fundamental interaction? Is there complete loneliness and isolation? Can loneliness be broken as many times as one wishes and still remain the same, elementary and pure? Is it possible to know what is not related to anything? Is the part hiding a complete whole, a homunculus of its own compound nature? Is loneliness hiding a homunculus from itself? Is the universe only a universe of interactions? Is the body of a dog or a horse only an interrelated set of cells, only a cell interaction? At the same time, are cells only the interaction of membrane, nucleus, ribosome, mitochondria, cytoplasm, and so on? And before going along this avenue, what is the minimum or simple interaction? Is it the particle and field interaction? Is it you and me, they and us? Is it the ballasted string that binds the parts like an adhesive tape or is it the decompressed spring that separates them? Is it the second law of inevitable perishing? Is it the first lyric law of love? Is all interaction love and attraction? Is the repulsion force of electric charges of the same sign, in turn, a resistance to separation of something that loves harder? Then, is the repulsion force a sort of attraction force which is reflected in the broken mirror of confused observation?

Let's go back to the search of an origin for life with this aforementioned implicit duality, kept in the pocket like a coin, now looking for any sort of information about it, looking for the father without a father or for the son without a grandfather; then, with the aid of possible technology and actual imagination, a minuscule multifunctional propelled and remotely controlled microscope could be injected inside a tiny piece of organic tissue. An appliance made for examining an ideal cell, custom-made for the experiment, a theoretical guinea pig-cell. Hence, once deep inside the flesh fibres and after crossing the lipid membrane of the cell in question, opened with a built-in minimalist kitchen knife, this mechanic traveller moves cautiously to the area of the cell nucleus, more interesting than the mitochondria and other organelles or cellular monuments shown in biological guidebooks. There, after taking a couple of panoramic snapshots, it makes another cut as thin as the flagellum of a bacterium and goes deep inside again. Then, in the interior of the nucleus of this eukaryote cell, the tourist goes for a walk until it can smell a chromosome near it. It approaches and begins to touch this biological object hesitantly, with the help of two skilled nano-hands. But, are language and classical laws still useful at this scale? Is the observing researcher inevitably and significantly affecting what he observes, staining everything with his dirty nano-hands? What is at the bottom of the ladder, what would be seen with the most powerful zoom? Is there a pixel of the entire observed object again, a pixel of the homunculus smiling at the camera? At least, is there a single photon imitating and representing it? And beyond, where mechanical naked eyes are useless, is the observer looking at himself? Does he appear in a mirror increasingly flatter and smoother, more and more faithful? And at the end, does the back of the eye ball that is looking through the microscope appear, like a traveller who had gone round the Earth faster than its shadow? Or does a new reality appear turning the corner, behind the mirror? Maybe, an unknown reality with new physical dimensions and laws? A new immense piece of the universe, more orderly and beautiful than the piece that is known now? A new universe behind the tiniest window, behind the shape of the smallest thing or the simplest process? A new part of the universe, expanding and taking place again until it becomes infinite? Just like a universe appearing behind a grain of sand that was blocking the neck of an hourglass, with two infinite conic glass bulbs? Otherwise, is there also an unobservable universe in the microscopic scale as in the case of the macroscopic scale? Something analogous to the part of the universe that emits light but that is so far away that can never get here? Something with such a small pair of legs that it would take more than a world to walk to the pupils or any other device used by the observer? Is the unobservable universe all the blackness of the night sky?

Working on the depth of small things, still with the eye on the microscope, now lighting and focusing on a certain scale, the magnified image made by some of the collected photons that leave the surface of a part of the mentioned chromosome can be seen now. Thus, touching and scrutinizing carefully what is seen, it can be certified that, at least in this ideal case, one is dealing with condensed chromatin, a sort of packed mass. And, after patiently unfolding this relatively compact thing, the anonymous mass reveals itself mostly consisting of a very characteristic and repetitive structure, the nucleosome, histone protein cores wrapped by large molecules of DNA; which is, by two antiparallel deoxyribonucleic acid strands, spatially arranged as single double helixes linked by hydrogen bonds – specific biomolecules that follow these and many other peculiar configuration rules and features, some of them ascertained by biochemists and geneticists over time, step by step, probably through very tedious methods. So, once separated from the histone spool and holding DNA in the nano-hands as if it were a treasure of gold necklaces and diamond jewellery, this brilliant architecture seems relatively magnificent and a more thorough and definite study is required. Accordingly, before releasing it, perhaps one must consider concentrating on its form and composition only. To do so, one has to forget all the things that are not important at this time, mainly what this biomolecule makes possible, that is, the genetic products derived from using the information contained therein, and therefore also the consequences of these genetic products; for example, what is happening now!

In this way, one watches its traits closely as one who observes only the raw material which statues or sculptures are made of, only the smell of the paint that evaporates but not what the author really wanted to show, probably the final expression that caused the artistic transformation of the raw material, the reason or the purpose behind it, the sense or the beauty of the exposed artwork. So, observing with the natural naked eye, can this biomolecule be related to any previous and alternative genotype? Can it be related to some source that also contains information and finally expresses it? Or is it just a copy of a copy in a sort of indefinite endogamic process until it reaches the spontaneous creation from nothingness or the creative reduction to nothingness? Is there any cosmic information? Or at least, any kind of information field blooming and expressing specifically biological information? In short, is there non-genetic information expressing genetic information? The pre-genes of the known genes, with other physical support, also part of life but not yet identified in the unstoppable process of acquiring and enhancing knowledge? Could this information be without any physical support? Does the word *physical* have any true meaning beyond the limited scope of scientific knowledge and the relative sensitivity and accuracy of the observer? Is physics helping to define a part of the reality related to another, the non-physical, the non-material? Is this metaphysics so to speak? Is the nature of the relationship or of the interaction what is sought in metaphysics hereafter? Or is it perhaps sense and beauty, what the picture shows in addition to paint? Is the metaphysics hereafter only the abstract expression of physical things? That is, is metaphysics the abstract phenotype of its factual genotype? On the other hand, is the abstract genotype of its incarnate phenotype? Is biological life the only physical thing really expressed? Otherwise, is life the only physical expressive thing? Does a stone not express anything? How about a sculpture made by a good artist?

Still in the laboratory inside a cell, with more and more imagination and less and less scientific precision, the minuscule microscope illuminates and focuses its parts again. There, once left far behind the perspective point where the irresistible beauty of a young and healthy, proportioned, symmetrical and smiling human figure is lost in the vastness of space, even lost within his or her own space, the deep facets show a view of a very small corner, where chemistry feels at home. Then, with a small mass spectrometer placed inside the nostrils of this pilgrim and fantastic microscope, the chemical composition of some indigenous samples is analysed. And after getting results and interpreting them, it seems that such a place shows a cyclical logic; indeed, there is an apparent vicious circle between the detected organic molecules that makes up the cell structure and functioning. Conveying it in a careless manner, from an amateur's point of view, like blind eyes beginning to feel an object by means of touching it with the cornea: the proteins that are part of the basic cell structure are formed from amino acids, but the amino acids are synthesized based on the codons of the genetic code – that is, in function of nucleic acids – and nucleic acids are synthesized through reactions catalysed by proteins. In conclusion, in this compressed instance, the protein seems necessary for the gene to happen, and the gene is both the source of the amino acid sequence and the protein designer. Therefore, the apparent occurrence of a first original biomolecule cannot be immediately or trivially stated. And this is even regarded as a paradox, *Orgel's and other* paradox; a paradox, at least, in this first glance that does not attach a specific weight to any prior consideration.

Hence, some daring arguments can destroy the foundations of this apparent paradox and can be used as tools for untangling the evolutionary history ball, woven like a rambling spider web. In this way, on the one hand, perhaps one finds the initial hypothetical biomolecule, aged in one of the extremes of the silk. Or on the other hand, after getting tired of clearing the intertwined and repeated branches of a thick jungle with a machete, one gives it up altogether, and among other possible alternatives, embraces the spontaneous formation of an organized complex group of interdependent parts or the absolute naturalness of this form of manifestation in cycles or closed loops; then, once the lineal logic and other wild obstacles have given in, it may become a biological axiom.

Thus, firstly, in a first machete blow, one can state a hypothetical paradigm that arrogantly assumes its validity, at least as a trend: *further back in time, simplest are life forms.* Didn't the bacterium come before the elephant? Wasn't it hydrogen before gold? A lone single proton before 79 protons joined together? Wasn't hydrogen the first element to appear in the temporary universe? Isn't a universe with only the hydrogen element more primitive than a universe that consists of hydrogen and gold too? However, if one follows the thread of evolutionary complexity reduction in an unbiased manner, doesn't it lead to the nothingness of absolute simplicity? Does it lead to nothingness or to an indivisible unit? That is, to something like a beloved son for his real mother? Maybe, to something that looks like a gravitational singularity? Just like a singularity with all the energy of the universe condensed and silently screaming inside the tiny and dark oven of hell: "I suffer from claustrophobia, let me out of here, noooooooow!"? And there, with the intention of further simplification, roasted on the embers of hell whilst seeing space and time vanishing and the guillotine coming down, can one assert without a shadow of a doubt that the simpler round egg came before the chicken with two legs, multi-coloured feathers and big orange eyes? But, is an egg obviously simpler than a chicken, even though it is not a chicken egg? Is a newborn baby simpler than his parents? If so, does the child tend to be more complex than his parents but always from a step behind them? Is simplicity a footprint of time, of superposed times advancing one after each other like an insect that is casting off a part of its body in moulting? Was the Big Bang the brutal eclosion of an egg? Was it a hatched or brooded egg? If so, is the known universe the continuation of the sound of the scream that broke the hell shell? Or could it be the continuation of the scream of the placental mammal while giving natural birth, while going into labour with pain from paradise to hell?

At another level, is any kind of universe the prolongation of the tiny squeak coming out of the egg when the embryo arises from a selfish Big Bang?

Conversely, now following the ascendant thread of complexity, what is the most complex thing today? Can it be understood? Is there a more complex ear for the more complex tongue? Are they, ear and tongue, always walking hand in hand? Does the current brain spend its time at the peak of complexity? What is the most complex brain in the universe now? Is it the brain that has a higher encephalic coefficient? What brain will in the future take the throne of complexity? Will it fit inside a skull, inside a bone box? Will its remembrances fit in the best of memories, in the most powerful of computers? Will it be a unique and lonely brain, naturally unaccompanied? Will this brain itself be an indefinite set of indivisible and absolutely solitary units, infinitely separated and completely isolated post-neurons arranged devoid of force in an abstract setting? Is the future of the brain to be stopped as if in a photo, the entropy that describes it as asleep and sprawled at the minimum energy level, at the minimum temperature? A universe of frozen bosons and fermions, silent crystal statues, information being ignorance that loves itself? No changes, an infinite sheet or field extended without wrinkles or folds, without forms, senses or any kind of pattern? So, in the end, will the brain be in the realm of the Third Law of Thermodynamics? After the last complex chicken the last simple egg, broken irreversibly into a thousand pieces? Maybe, the last egg because it is so simple that it cannot do a thing, just a dot? Could this end simply be an unfertilized egg that will never hatch? Is the maximum complexity also perhaps simplicity? Is moving forward to complexity also returning to the origin, to absolute simplicity? Isn't progress making it all easier, making a tool that automatically solves the problem? Even so, is it necessary to increase brain complexity to simplify problems? Is this a contradiction, a migraine or a headache?

Travelling through a less apocalyptic route, between the Big Bang and the Big Freeze pavements or other helpful concepts – at least helpful for some of the Holocene Quaternary peers – that describe alternative beginnings and endings, such as the Big Bounce or the Big Crunch, the evolutionary biological science also seems to take the downward thread of complexity in search of the origin of life. However, climbing down the well, it decides to stand just before the horizon where the Sun of the past is rising and setting. That is, the hunt stops at a very little old village of the past, but still a site with some complexity, more like a small city. A marvellous place in time, terrific space-time where now is stated, billions of years later, that something unique is observed, maybe a biological singularity. Specifically, the beginning of a primitive small cell, which was probably the common ancestor of all current cells, its descendants; perhaps, an asexual prokaryotic cell with a genetic code based on DNA, coding for the formation of amino acids. Additionally, also presenting some kind of membrane to protect itself from the environment, probably a lipid membrane surrounded by liquid water with some simple atoms and molecules dissolved in it; surrounding water which was, at the same time, an essential part of itself. The first or the last universal common ancestor, as it is commonly referred to depending on the point of view.

In short, there was little new under the sun; billions of years earlier than current organisms and words, the academic definition or description of cellular life, as an uncompromising obstacle, remains intact embedded and adapted to that past epoch too. Moreover, there are such quantities of accumulated evidences supporting the same in the subsequent period that they form entire calcium carbonate islands and mountains distributed throughout much of the crust of the planet; the same mountains where people climb for fun or build houses and castles that are made of the same material. Anyway, this is obviously a sufficiently complex definition to neglect or reject the curiosity about the precious constitutional process of this virtuoso, so delicate and valuable single object named here *common ancestor*. The father and mother of every living being and therefore, able to replicate and produce sons and daughters, new living beings, indefinitely; that is to say, being able to give birth to incestuous descendants that have become more and more complex, diversely complex, even though to the point of forgetting the simplicity of their parts, that attach them indissolubly and nostalgically to the past, to that ancestor that seems to perpetuate itself too. So, is biological evolution this upward thread that preserves basic attributes whilst secondary attributes are being removed and added? What has never been able to stop or cut the wall or the scissors of natural or artificial selection? Why not, was it too basic? What is the present continuous of living beings, is it life itself? What living thing of life history has always avoided the precipice of death? Perhaps, some of those genetic structures identified as dead or lifeless organisms? Are viruses and plasmids mortal? Are viruses and plasmids immortal? Is there anything amortal? What would be the full living phenotype of a living virus? Is the one that meets the desire of its lack, a home where it can reproduce itself, to recur again and multiply? Is it the infected body of a host animal that uses its own cell machinery to unconsciously reproduce it? Then, is life of the virus a man with a cold? Then, is the death of the virus a man without mucus?

Under the shelter of prehistoric and historic facts, and once realized the personal evolution of the common ancestor – so many times pruned when it entered too dangerous places, too fragile floors, unknown cul-de-sacs, unbreathable zones, etcetera – in order to define life in biological terms it has also become imperative to support it obviously in the *reproduction* and also to state the aforementioned tendency to *indefinite complexity*. Hence, one can imagine an information flow, an evolution flow following a tangled direction, flowing down the rocky rapids of a meandered river or running through a bumpy road with roundabouts, with sudden and steep ascents, with circumstantial swings, backwards and forwards, but always retaining also some of the luggage and furniture of the parents while tending towards the flying modernity by means of more complex sons, with more and more functions and capabilities; towards more domineering living beings, with more layers, like a branched tree growing concentrically, as present itself that swells up and spreads out as time goes by. In short, the evolution of life understood as a whole that residually accumulates while achieving a higher environmental dominion; although in this way, being also capable of folding like an accordion or refolding like a protein, obtaining deeper and deeper dominance, until mastering itself too. Thus, being also capable of taking orders and obeying them, whether willingly or crying, snivelling or protesting, either through obedience to legal or wild imperatives; or for whatever other imperative that will be exactly the same petty rubbish for the innocent and free.

Indeed, as life dominates or tries to dominate the surroundings, certain living parts also dominate other ones; perhaps, in a scaled or weighted manner. For example, one can observe, from a slight dominance of one allele over another in a diploid cell, up to the most brutal forms of domination, as a lioness sinking its retractable claws into the back of a young impala meanwhile also biting and closing its jugular, strangling it, to end up eating it raw with the family before digesting and expelling the useless remains while fertilizing the porous soil of the savannah. In other words, trillions of cells arranged in a lioness, as part of a fearsome criminal supraorganization of lions, hunting and putting down trillions of hierarchal organized cells, as in the best of armies, in an impala, their chief or supreme leader, the spirit of the people. All of them, parts and assemblies, sovereigns and subjects, descending from the common ancestor. So, if it were watching this show, what would the ancestor think about it? Which of them, lion or impala, look more like the father? In which of them would it recognize itself quickly? Both equally, two ripe fruit of its personal evolution? In the quarrel itself? Is it a lovers' quarrel? In any case, is evolution of living beings over time really the evolution of a single living being, of the common ancestor? Then, are the current living beings also the evolved and evolving ancestor? Is evolution a concatenation of brothers or sisters? When did the common ancestor become a family with all kinds of affiliations and enmities? What kept the family together after cloning and splitting? Is the evolved ancestor a family playing solitaire? Otherwise, who could be its opponent, a bored stone? Is complexity a sign of triumph? Is the domain feature, the hunting lioness, the best sign for life itself? Or is life better represented by the impala? Is the sacrificed lamb the most basic expression that builds everything on its back, such as the alphabet? But, does the impala sacrifice itself in an act of love? Does it let itself be caught in a last imperceptible and compassionate moment that does not hurt the lion pride?

Now, let us go back into the waters of time and simplicity, looking for simple hidden treasures, surely lost in the abyss; there, it is claimed that one can observe, before the common ancestor, something on Earth — or wherever it was — that was not a cell yet. That is, an amalgam of organic molecules, something that looked like an incipient gene, a teen ribosome, an adolescent mitochondrion, a baby lipid membrane or a mixture of any similar undefined cell piece. Perhaps, a very simple virus or a plasmid, things like this intending or pretending to be a cell, like children who were playing at being adults, satisfied while ignoring what is expected from them. In short, an innocent protocell, and therefore, devoid of original sin and karma, still unborn to the world of acts — this polluted place that will have to control, to load or to put aside. Nevertheless, although it was not a cell yet, what was missing for the presence of life? What did this protocell need to become something alive in the matter womb? Had this protocell not replicated itself yet? Does a stone replicate itself as it goes from one second to another? Does anything make this effort in its place? Hasn't the universe, including the protocell and the stone, evolved from more simple forms? When did the stone become a stone, when did the change become strictly evolution? When did the change diverge and disguise itself? Isn't a stone more complex than a grain of sand or an electron? Is number four more complex than number two? Isn't the multiplication more complex than the sum, two times two more complex than two plus two? Can a problem be more or less complex than its outcome? And, can the outcome be more or less complex than the problem set? Before the common ancestral living cell, wasn't there a skilful driver at the wheel of the vehicle, wasn't there wilful autonomy and mastery of matter and energy? Was there a driver but not a vehicle? Didn't the vehicle start at the time, lacking pieces or petrol? Are there pilots beyond the laws of physics, beyond determinism, with or without the cell being? Is the law a pilot who does not stop or turn when there are bends? Do natural laws run without the need of petrol? Is there any freedom in this land? Is a stone free, without rules to obey or needs to satisfy? Or is it the symbolic physical expression of absolute slavery, the perfect sign of a slave? Is it the beloved son of laws, always paying attention and honouring its parents? Likewise, are laws free, are they outlaws?

Within this framework, although without solving every or any of these questions and encouraging part of the queries again, the final constitution process of the cell is often called *abiogenesis* and its current studies aim to show the series of chemical reactions, the sequential reactions between inorganic and organic molecules which eventually led to the biomolecules that would constitute the first basic functional structure of the common ancestor. From which and with which, as previously mentioned, the known biogenesis began, to endure until now, an unbroken line of descent of pureblood, aristocratic. Therefore, abiogenesis is trying to describe the origin of life as the appearance or emergence of a predefined object, sustaining it on certain established parameters, with the assistance of the oldest fossils and the theoretical science derived from them, also with the knowledge of the geological and the past climatic conditions and the endless hope that the foot of the common ancestor, or a very similar cell or thing, remains not far from the glass slipper under study. However, is the common ancestor the parent without a parent? Is it the first parent and full stop? Was it the full beginning? Was it the father and the elder brother of its horizontal clones, the elder brother and father of its vertical sons? Was it, if telling the story backwards or forwards, only the first or the last common father or mother? If we go further back from this common ancestor, is everything a parent or is nothing a parent? Was it the first genotype that expressed some information, like someone dumb or a baby that starts talking all of a sudden without having heard any words before? Was it the first intelligible phenotype? Was it the first shaking of hands between the genotype and the phenotype? A mystical moment of physical and spiritual communion? Something like the emergence of free and shared love? Or was it rather the success of a sentence, an inflicted punishment? Could it be something aseptic, another typical day stuck in a rut? Can anything go faster than light and go back and know what the past was exactly like, what the origin of life was like? Is the act of remembering a valid proof that some neuronal processes exceed these limits? Is remembering the act of a mind that is draining? Is it possible to remember the days before birth? And the day before conception? Is conception a firm and solid concept, is it pointing to reality with its finger? Is it just a concept? What came before the concept or the conception? What is simpler?

In order to deal with the origin of life and, likewise, to be able to describe it accurately, the study of abiogenesis has generated several hypotheses. By way of example, some of which lie within one of only two simplified movements, the so-called *metabolism based via*, or on the other hand, the so-called *replication based via*. These assumptions base the primary distinguishing features of living beings fully on metabolism and replication, thus describing them as the pillars of cellular structures and activity; therefore, around one of them, as a sort of condensation nuclei – like a pearl growing inside the shells of an oyster – the definitive cell formation happened. Perhaps, it can also be said, the common ancestor happened or took place. The rest of the baggage would be given by later additions and subtractions, slowly moving two steps forward and one back; and, if abiogenesis walks along the same path as some academic and ideal biological evolutionary theory, for instance the *punctuated equilibrium*, the protocell transformed gradually but through *stasis* or dead calm spaced in geological time. That is, by means of quick and triumphant changes taking place followed by long periods of invariance. Like kicking, like playing golf, stroke after stroke while using most of the time to find and get back to the ball again. Incidentally, is punctuated equilibrium a digital evolution theory with a quantic look and quantic rhythm? Is life evolving from stroke to stroke, from hole to hole, from quantum to quantum? Has the ball of evolution ever dropped into a sand bunker or a deep lake? Has it ever gone simultaneously into multiple holes? Might it be possible to enter it in one stroke alone, a hole-in-one in a par-five, a condor? Does the club change depending on distance, wind strength or rough surface? However, is there a hole where putting the ball or the hypothetical golfer of evolution will play indefinitely? Would a golfer play golf without holes where holing out the ball? Is this the reason why so few play this boring game or any other indefinite game? Is there anybody that inevitably plays it? If so, is it just to see who throws the ball higher and farther, over and over again? Then, does one win when the ball, as a satellite or a space probe, finishes orbiting the Earth or escapes from its dominion and leaves the Solar System dominion behind too, as if it were the Voyager 1?

Thus, without becoming obsessed with rigor and not worrying about substantial differences or confusing mixes, in the case of the abiogenesis hypothesis based on metabolism, a particular process of incorporation is initially described by grouping several plausible organic molecules in a certain old planetary environment. That is, chemical entities driven by means of gravitational and electromagnetic interactions which finally led, because of the mass and electronic nature of these molecules, to a chemical system carrying out self-catalytic processes. Then, perhaps the organic coterie started a drifting journey immersed in some kind of planktonic flow; in this case, this mob gradually increased its structural complexity until a sort of band was formed, an organization increasingly similar to the target cell; especially, when a selective membrane was built or consolidated around it, allowing gradual specialization, mechanization, rudimentary patriotism or nationalism, already believing its own lies. In a few words, one small social community dawning and glimpsing the rest of the world and in synchrony, also glimpsing at itself; that is to say, beginning to suffer from division, smelling the I, suspecting the ego, predicting the self. In the most radical point of this hypothesis, any pure mechanism of replication was still neither present nor necessary, just something like a stepwise chemical mechanism on some kind of support, like a cloud that is shaped in the blue sky by a group of tiny water particles around one tiny speck of crystalline solid powder in suspension.

On the other hand, in the case of the hypothesis based on replication, studies are focused mainly on the RNA molecule, making up a whole world from RNA; establishing the more consistent mechanisms of its formation, because once formed, its special chemical nature allows, as it is widely known, both for storage – as seen today in some viruses – and the enzymatic catalysis. In conclusion, the plausible natural path genesis is studied, driven and helped by the same natural forces mentioned, of the molecule that could merge into a single entity the current role of proteins and DNA.

Anyway, for the support of these and other theoretical hypotheses, practical laboratory experiments are carried out too. In some of them, the causal formation of organic molecules such as amino acids, exclusively from inorganic molecules present on the planet early on, has been demonstrated for some time, manipulating reagents under specific conditions which imitate some possible environmental scenario billions of years ago. For example, ignoring a great deal of it, something like mixing water with other specific inorganic elements or molecules, such as natural catalysts or dissolved molecular gases such as nitrogen or carbon dioxide, and cooking the mixture at different temperatures, love and pressure conditions; meanwhile, exposing it to specific light frequencies, electric shocks or magnetic fields, and so on.

Also, other practical experimentations, up-to-date experiments but still following similar routes, attempt to synthesize directly the organic molecule that once present can replicate itself, spontaneously or when changing the culture conditions. Obviously, if there are enough essential raw materials in the environment to do so. Basically, the conditioned synthesis of RNA or very similar biopolymers obtained from inorganic stuff.

Otherwise, as a very simplified note about other speculative trends, the possible biological replication that would turn out to be from crystal growth is being theoretically and experimentally investigated too, since crystals show new structural patterns arranged on previous crystal patterns already formed, similarly to the way of genetics. Thus, some crystals of certain characteristics, by means of degradation phenomena or chained chemical reactions, perhaps became the first biomolecules that sustain and constitute the current cellular life.

Well, it is easy to put these trends within a frame, despite the fact that the mentioned abiogenesis paths are not specifying dramatic facts or circumstances, seeming far removed from common feelings or emotions. As a result, both in the chemical model of the autonomous system with metabolic activity and in the model of the sole molecule-world capable of replicating indefinitely and also developing more complex molecules and worlds, it is observed how on trying to approach the common ancestor, the abiogenesis hypotheses argue from a certain point of physical complexity, basically molecules with a specific degree of binding. Chemical entities that already have some size, some stability; and therefore, they are not moved due to a breeze, light thermal currents or critical or favourable opinions to their proposals. So, it is usually accepted that the core study of theoretical and especially experimental abiogenesis research does not deal with other little subunits nor imply its previous formation. In other words, at least the isolated particles are considered to be orphans and extremely wild people, although they have ended up collaborating submissively like pet lambs. Besides, it should be noted or be inferred from what has been said, that the gaze of abiogenesis speculations is not often placed in synchrony or near the beginning of the known universe.

Hence, among many other considerations that can be alleged to the theoretical and practical field of abiogenesis, it seems that time spent since the Big Bang to the last moments before the common ancestor is part of a process that is considered impossible to imitate, a lost cause, a complex sequence impossible to be traced, defined or represented with precision, second by second and step by step. This wide interval is commonly offered all at once, just like a cake with one or two garlands, baked by natural laws and mother Earth, almost irrelevant to the subject of study; despite belonging to the real historical formation of a particular structure as well, animate or inanimate, under the hypothetical guardianship of the same natural laws. And, it is also accepted that the role of human beings, the artificiality of the facts, can be ignored or deleted from the experiment design and its results, but also from the elaboration of the assumptions and the understanding or validation of the theories.

So, what one does in the laboratory building or in the mind lab, can be fully expendable, subtracted without any fear from the crucial point which would represent the beginning of life from inorganic sources or from a non-biochemical organic source. This artificial event would also occur or could occur without human beings and their minds, in the middle of an inhabited wasteland desert, when what has artificially been made is naturally given. However, is there anything wrong with all of this, with this research or interpretation method? Is there any difference between artificiality and naturalness? Are living beings and their living cells natural beings or artificial beings? Is the pluricellular organism the artificial being of cells? Is life the artificial phenomenon of a natural universe? Is there abstract artificiality or is it only a concept inevitably attached to human participation? Is nature natural or as Salvador Dalí himself precisely said: *nature is supernatural*? Is there something that can be defined as subnatural, supernatural or anatural? What exactly do human beings do in their mechanized labs, what is their main role? Will human beings demonstrate in their laboratories the need for a universal creator for life to emerge, or will they die in their attempt? If they become successful as creators, will they also bring up the creation? Is there any difference between *bringing up* and *creating* in this case?

Furthermore, as is well known, surprisingly this last step – the natural formation of life without prior living existence – is not usually said to be observed now, beyond some very specific contexts that often lead to a quick and categorical response. So, one cannot say aloud that life is emerging or is forming from scratch now, here and there, around the corner, every few minutes. Also, this subject does not seem to be a point of investigation or research carried out by theoretical or experimental abiogenesis. It is a sort of biological axiom that a single life covers the whole planet and all the living beings have the same ancestor, the same flux dissolves them in the heart of the past. From this point of view, nothing new is formed now as this common ancestor or other possible ancestors formed yesterday, although perhaps the primitive conditions should be very similar to current conditions to already allow the delicate cell life; obviously, very similar in those things that life itself has not changed while packing and unpacking the Earth since then. Thus, the question of the origin of life seems to be complex and with a lot of implications, and some kind of original formation from inorganic sources on the current Earth planet is far less expected than an extra-terrestrial or alien holiday visit from outer space.

That being said, it seems that life itself does not like competition, as it is shown through its own voracity towards itself. No one would have ever observed a living thing forming from something that is not alive, neither naturally nor still in the laboratory; meaning by that, with human beings acting as artisans, breeders, inventors, and so on. And, as long as the hand and the ingenuity of present living beings participate in this process effectively, touching with their living fingertips the fingertips of inert or dead matter, sowing prefabricated seeds or modelling new living beings with ink, brain or clay, it seems that it won't be possible to say it while being right. Or on the contrary, would it be possible to say it with absolute certainty? What is needed to observe this special process of emergency, the abiogenesis process? Is it enough to have a son being a father? Or rather, being a father while being nothing, then observing from nothing the beginning of something? But, can one be a father or a mother while being nothing? At least, to be a parent and absolutely nothing else? Anyway, is it enough to observe the emergency of filiation, the father becoming a father as the son becomes a son? Is it enough to have or have had a father, to be a son or a daughter? Indeed, is it enough to emerge from scratch, from zero, from the darkest and deepest abyss of the flesh and the past? To arise from the open cone of the past once reduced to the instant of its present vertex and then to begin? And then, to be born and cry after being kindly patted? But, does the newborn baby welcome everybody to its own life? Does the newborn say hello, I am here, I have no memory and no history, with no remembrances, what the hell is this? Is being born something other than being out of the womb? Is the beginning of life being born? Is the beginning of life a hydrogen bond between two DNA molecules, a chemical bond? Is the beginning of life an electromagnetic attraction between atoms or among stacked molecules with or without an owner? A baptism, a vital record? A bit of pollen in the wind and flowers in bloom catching it while rhythmically dancing with the same wind? A drop of water on a lettuce seed? Bags of pheromones entering the nostrils and travelling to the brain, a breath of life? A look of love between two apes? Then, is interrupting a look of love or any sexual attraction, whether it has been unilaterally or simultaneously expressed, also to annihilate life to its last conclusions? Is the hydrogen bond bridge stronger and more solid than the bond bridge built by a look, by desire, by will, by chance? By affection, by care? Is turning down the sexual relationship also an abortion? Is cutting any kind of relationship or physical interaction also ending life? What kind of interaction or relationship is always life without an ending, no matter what bond is untied or bridge destroyed? Is life inseparable from living beings? Is life the relationship between living things, the creative addition or the synergy emerged from the absolutely unique, free and individual living beings? Is it rather the direct and personal relationship of the living being community and every living being alone with the dead environment, face to face? But, is life ever individual, is it ever particular life? Is life something more than individual and non-transferrable life? Is life just a cut that hurts, that one feels? So, is the universe or is life broken up in parts?

_3_

The future is the skin of a heart, of a beating present.

Ψ

THE FUTURE,

the phenotype and the singular automaton

Inspired or not in abiogenesis and biological evolution, many of the most recent initiatives and cutting-edge biotechnology activities involve the transformation of basic pieces of information and consequently also the transformation of the most essential biochemical processes. Basically, this practice is sustained and developed, among all kinds of other institutions, by the main actors who are also involved in biomechanical and medical progress, as well as by those who design industrial production processes or optimize the existing ones; especially, in the field of pharmaceuticals, agriculture, farming, food processing and other similar businesses. That is to say, the enterprises or initiatives with some sort of contact with the biological world and its possible benefits. In turn, much of the research, investigation and development are carried out in academia or university, expanding and building up more theoretical knowledge, opening new doors for active experimentation on every front. Finally, knowledge is feedback from the practical or technological activity, and in this way, this cyclical process is pointed towards the unexplored and the unknown.

But, does anyone know what the unknown is like? Where is it and when does it arrive? How can one go to an unknown place? Just by closing one's eyes and covering one's ears? Can one recognize something that is not known in advance? Indeed, can one recognize unknown things? Is one unconsciously and completely surrounded by unknown things? Is the world a continuous interpretation of the same known world, a continuous updating process? Is it just a reinterpretation that happens when the observer changes, then varying the perception of a constant universe? Or maybe, does the observer remain always constant, everlasting while the moon waxes and wanes? Is it always the same clever engineer, both for the wood abacus and for the quantum computer? Is it constantly the same wild attacker, both punching and stoning a hundred thousand years ago or spreading mustard gas and throwing the atomic bomb from a plane not long ago? Equally, no matter the amount of art made, is the artist always playing with the same clay? Nothing essentially new to observe or no one who observes it for the first time? Isn't it possible to lose one's way and not find it again? To lose one's way and prosper in the unknown?

Although the unknown does not leave any clues or evidence, it is known that the biotechnological innovation, which accumulates a vast knowledge and a lot of work done, performs the manipulation of genetic material and consequently of primary genetic information stored therein, to be finally expressed in a different way. That is, it is modifying the genomes and therefore also those associated phenotypes, perhaps some of them still common today. Basically, by means of genetic engineering techniques, using the conceptual theory of this science and its technology. Thus, the artificial mutation of the present genes on the planet is accomplished over any species, such as those genes of the exploited vinegar fly. In short, genetic material is processed and sown, modified and planted directly in the field of a host genome and, if the seed takes root in the soil, the host grows according to a new genotype, as a modern mestizo, half natural and half artificial, half a mule and half a green horse. But, what will emerge from flat shadows, will some kind of racism emerge towards the artificial half? Or will a new form of racism perhaps emerge towards nature? Anyway, are they the same in this process those who sow and those who reap the harvest? What does the vinegar fly think about it? Is it still the vinegar fly? Is it the same to plant as to root? Are sowing and irrigating sufficient steps so that the seeds can finally take root? Is the one who plants a gene also the one who roots it in the host soil? Who or what ties the ribbons between wilful roots and indifferent soils? Can it be simplified to the most basic natural force or interaction?

To address and better understand the nature of this field of study and work, it should be noted that these artificial genetic changes may seem somewhat very intimate, manipulating on a deep, small and intricate scale, precisely and powerfully – able, if one should want to, to improve the sight of an eye that is designed to grow on the nape or to make a rounder and firmer bottom that is forced to grow on top of the head – but, these are processes of the same primary nature as grafting a branch of a plum tree to an almond rootstock or choosing a stallion to copulate with all farm females while repelling other males with the noise of the whip or the sharp touch of the spear; very much like participating in the conscious choice of a sexual partner to form any type of blood family, whether participating as a spontaneous lead actor in a simultaneous selection based on sensual or intellectual attraction or as part of a cold and calculated relationship based on tribal law or enforced by a matchmaker; likewise, if relationships are left at the hands of fate to randomly recombine or when one is choosing and matching for any reason, both personal and universal, both rational or on the verge of individual or collective madness.

Hence, without any exclusiveness reserved to modern ages, current genetic engineering walks along the same path that has been trodden since immemorial times. In spite of this, obviously, it would be wrong to deny that nowadays there has been a huge breakthrough, a big positive change in traffic speed. But, does it mean that the road has become smoother? Is the way being paved? Have the shoes been changed by modern, adapted or tied ones? Or is the road that is accelerating under the feet of living beings, which have to run faster and faster as in a runaway treadmill? Does the runner need brakes, could something be run over? Is it inevitable to run off at the next bend or leave one's teeth sunk into the ground when the legs cannot keep up? Does one need a seat belt? Is there any dangerous thing on the street on the lookout to stop the runner abruptly and all of a sudden? Are there traffic lights and police, any intelligible traffic rules? Is everything following the natural course despite radically changing the intimacies of genes, being the artisan another natural piece of nature? Isn't it natural that biological species squeeze all their possibilities with the tools at their disposal? Isn't life squeezing its environment? Squeezing all the available resources so it can spread like a forest, isn't it the nature of nature? That is, to spread over the continents, the oceans and also other planets in other star systems, isn't it the nature of life? As simple as this, to occupy space and time, to be a growing space-time event, isn't this the nature of the tree? To grow until sustainability or stability limits, until it is brought to a stop, until it has run out of suitable energy?

Anyway, whatever artificial or natural genetics, these are included in craft matters; that is to say, in this context, informative processes and the subsequent coherent making. In a few words, formation derived from certain information. Therefore, dealing with genetics and progress or evolution, is it possible to construct a radically new artificial gene?

Can genetics contain imagination? For example, is it possible to manufacture a genotype that once sown and embedded in the host, as a plant in the substrate, will it develop exceptional new organic traits, qualitatively different phenotypes? Iron flowers with leaves of gelatinous gold, releasing cut and transparent diamond pollen? If so, what will these new features be? What are the most demanded and dreamed ones? How about some organic rockets on the soles of feet or heads with aesthetic pillows to rest anywhere? A touchscreen on the palm of a hand? Or even better, a screen developed jointly with the eye-brain system during pregnancy? Also, better and crazier imagination to go on pursuing something without getting bored? Are there any impossible features to develop with an effort, patience, ingenuity and available resources? What are the limits of the genetic code? Will new rules in the code be introduced? A new basis, a new grammar perhaps? Will somebody finally see an engineered living body staring at oneself without knowing what it is, as if one were listening to foreign languages, without understanding anything at all? Merely, will one be able to recognize that a language is heard, that somebody is talking? Beyond likeness, can a brain recognize a more complex life than itself? Is this the extra-terrestrial life that some people are searching for unsuccessfully? When will a completely new code be necessary due to the obsolescence of the current one? Is biotechnology following a process of trial and error, looking irresponsibly surprised at artificial or natural selection taking place over newborn genetic products? Or conversely, does it design and choose knowingly, to undoubtedly impose a dominant trait or making it useless for the sabotage of enemy land? What will the homeostatic response of all natural, global, galactic, universal hosts to the artificial modification of genes be? What was the response when fire was controlled or when human beings began to domesticate animals or sow the land? Is it still to come, is anybody waiting for it?

Besides, biotechnologies are not limited to the specific use of pre-existing genetic material, since they are making or soon will manufacture with masterly skill and expertise, also organic-mechanical life products; indeed, very complex handmade biomolecules constructed from non-genetic organic sources. Initially through imitation, and after that, when beginning to learn and mastering the techniques, incorporating the most original inventiveness to this process too. Thus, as a common simple chemical reaction, the composition of big biomolecules from much smaller and simple things than themselves is possible. For example, once the method has been worked out and the obstacle of time has been overcome, a gene and finally a synthetic genome could be made by hand from tiny molecules or single atoms put together, using them as primary building blocks and only following the acquired and limited knowledge as a precarious roadmap. Maybe, also with the help of fortune and nature – as if it were a network for the tightrope walker that covers some of the holes that unconscious ignorance inevitably left unattended. Then, imagining a very simplified sequential process, with the addition of the knowledge about proteomics and lipid envelope formation, it is possible to form a simple virus. However, will this virus be defined as life by its builders, would they consider themselves as bioconstructors? Will they quench the thirst of the virus of someone else's reproduction machinery as if it were a sexual vampire or will this need to be improved so it will be able to do it even at a distance? Will biotechnology endow its viruses with autonomous machinery for their self-replication and thus stop their annoying behaviour to the neighbourhood? Will the artificial virus be set free to satisfy its needs among particular enemies, among other body diseases or among country, sea or air pests? And then, will the bioconstruction process go forward indefinitely, towards increasingly complex forms? But, is it possible to build something more complex than the builder? Will there be time, intelligence, resources and patience to do so? Is some help needed to do this? Are zillions of bioconstructor brains needed to build only one more powerful single brain than any of the bioconstructor particular brains? If so, is it possible to build a sexual couple out of these new complex brains to allow them to start procreation? Is the help of someone else always needed to increase complexity? Then, is it possible to make up something more complex than the whole which is making it up?

When talking or speculating about technology goals that need atomic and molecular manipulation and about some of the means of making them possible, as well as clever brains and the powerful computer controlling the whole process, one has to also take into account the invaluable help of the most advanced nanotechnologies; for instance, the use of nano-tools or nano-robots with more or less autonomy and able to perform strenuous, delicate and tedious work at atomic scale, in the quantum mechanical and empty underworld. Consequently, to comply with the described progress, from atom to virus, from virus to cell and from cell to infinity and beyond, these tiny devices seem to be required, at least in one of the viable possibilities. Perhaps, nano-robots remotely working in aseptic construction zones or otherwise nano-explorers penetrating deep into the hostile space of a living flesh body; trying to learn and go unnoticed to armed native resistance, avoiding the setbacks of a dangerous metabolic world. Anyway, similar to the small multifunctional microscope with nano-hands presented before, though with a lot more work to do, these nano-devices will have to sort out every atom and every molecule according to a very complex blueprint; and if one does not want to lose patience and one's whole lifetime watching them working, it would have to be done by many of them, at a tremendous speed, as if they earned money for each atom or molecule placed in the right place. So, perhaps these fantastic tools will also need a high skilled mechanical body with a wonderful portable toolbox, also helped by their own supreme-like artificial intelligence based on an elephant memory programmed with precise knowledge about fundamental interactions;

for example, a genial automatic understanding in scientific fields such as electrodynamics or quantum thermodynamics or about kinds and forms of immunological responses; especially, to safely go into auto-protected territories of living beings. In any case, putting together atoms and molecules until a stable structure is formed as if by magic or until the threshold of the spontaneous formation of this structure is reached. And then, due to its physical or biochemical properties, inserted into a living body, it fulfils or immensely improves the biological functions, as does an ordinary prosthesis such as that of a hip or a gold false tooth, or as it already occurs with some of the non-sabotage processed genes planted in the body of the fruit fly.

After all, real or figurative nanotechnology at the service of bioconstructors; until suddenly, in a few geological moments, whatever the step or the bridge leading to the next floor or to the neighbouring country, the construction of a first whole prosthesis will become an actuality, something like an organic robot, with a face, a heart, hands and whatever the living beings of that moment have, including their knowledge and the skills derived from it. Maybe, leaving aside the simple and boring robotic cell, something much greater and fantastic, a biorobot or biobot able to play the guitar and sing opera always in tune, teaching physics and philosophy, ploughing the land, cooking exquisitely or cleaning the toilet until it shines like gold. A fleshy automaton resembling living beings, touched upon in novels, essays and cinematographic science fiction long ago; in some cases, knowing even less about theoretical knowledge and the appropriate technological tools to make it effective. That is, an organic robot in the image and likeness of what has been observed in natural life. And after that, imagining and inventing, perhaps also supplied and fitted with proper traits. However, will it ever be comparable to the entire body of a living being? Will it be possible to install a whole organic body as installing a prosthetic arm or leg? If so, will one need any kind of permission too, maybe to certify the lack of a member? Is the entire human body a member, a member of humanity? Will painless and quick amputations be carried out to make room for these new technologies? Is the healthy brain some sort of insurmountable barrier for the bioconstructors or just a very difficult obstacle, a much more complex art? So, when installing an entire working body, what will remain of the patients or clients? Will life remain without them, life without people? Will an old cell be left at least, a cell to be remembered, a remembering cell perhaps? An entire hair or a nail of the little toe instead? Thus, will it be a biorobot automaton living the life of a dead man? Anyway, will it also be possible to build genes or other cell structures from far below the atomic scale, to manipulate subatomic particles or strings, fields or empty space directly? Is it possible to build a yocto-robot as simple as a piece of mass or time but commissioned to build a nano-robot warrior? A son of war, yocto-sciences and biomechanics? A nano-robot to combat viruses and rebel cells from deep inside them, one by one and with the speed, strength and courage of millions of white blood cells? Will it brandish a nano-lightsaber or a big nano-flail for the sake to intimidation? Is it possible to build a flail with half a dozen atoms? And a five-fingered hand with four atoms?

And once one lets imagination run wild, using science fiction again and being confident in the future, how far will this process of construction go? Is it possible to build objects capable of sustaining themselves successfully at any expense, whatever the cost? Therefore, building objects which are able to replicate themselves, to reproduce and become diverse, to evolve slowly without realizing it or abruptly after stasis but indefinitely towards more complex forms? Then, will they be able to build more complex self-replicating objects doing the same thing and better as well? Is it possible to compute everything that is needed to accomplish this? Can one program this path, getting it started with a conducted shove? Are incomputable things an insurmountable obstacle? Are hindrances the limitations of arithmetic, any natural limits of computation itself? Has nature overcome all these limitations? Is there something incomputable under the skin while being necessary part of the orderly organic structure? Isn't there a universe with life subjected to any known or unknown limitations? Is existence blocked by what is incomplete or inconsistent in itself? If not, is this limited automaton already running? Will the era of computers and neuroscience be the era of transferring all brain qualities from flesh to another harder interface? Is it part of the same process that happened in the fields, from calloused hands to tuned tractors with radio and air conditioning? But, is everything computable in a living being? Is life an integer or an irrational number? Is it any other kind of number? Are the *countless* only ephemeral concepts of a present that is moving towards an undefined capability to count? Are relationships among living beings computable too? Are they predictable too? What about the vital synergy that emerges from the grouping of complex and unique individuals? Thus, is dynamic society as a whole computable or measurable? By whom or what? Is what is known just the surface of the computable part? Is the universe computable from head to toe? Is it possible to reproduce it entirely on a computer, including the same computing computer too? So, is reality absolutely representable despite its being scaled? Is it possible to simulate reality? Is it enough for it to articulate the word *universe* or *this*? Who or what will be able to compute and understand the calculations and words of the most powerful of evolved automatons of the future? Just another more powerful automaton, its own invention, its baby? An automaton made to discover its inventor or what is invention itself? That is, to know itself? An automaton made to know and represent the truth, a technological singularity? Anyway, will the revelation for this automaton come from beyond the event horizon, from the world of gods, from heaven or from the twilight zone? As the *Tanakh* or the prophecies of the *Old Testament*? As the disclosure of the *Quran*? As the revelation of the universal soul of Krishna to Arjuna before the fratricidal battle? If so, could this automaton play the role of Krishna as well as of Arjuna? However, is revelation the only way to know in a limited or finite world, in an incomplete and inconsistent reality, constantly in doubt and by default become ignorant? Definitely, can the programmers program a true revelation for themselves or for the program itself?

A great deal of study and research is currently underway in some transversely related scientific fields allowing us to make sense of some of these and so many other possible questions; matters based on biological and mathematical theory, computation and programming and much to do with the automata theory and practice. So, only as superficial related examples one may mention the Turing machines, the Von Newman universal constructors, the rapid reproductive prototyping, the three-dimensional printing, the compilers design; the entire fields of robotics, cybernetics, bio-mimetic mechanics, artificial intelligence and avatars; the study and analysis of formal grammars and related languages, biological mathematical models, neural networks, game and information theory, complex adaptive models, combinational and sequential systems and a huge etcetera that includes, or will include, all possible knowledge. Nevertheless, what is this technology really about? Is it anything other than a toy, a specific prosthesis, a weapon or a practical or theoretical tool? A medical toy, a psychological prosthesis, a heartless soldier-tool? A prosthesis for memory, answering as an intelligent encyclopaedia before being asked? A tool for mums and dads, entrusting it the traumatic natural childbirth and child entertainment? Anything other than a vehicle or a slave? But, is there always a need to ride or enslave someone? Is it a basic necessity of life? The capital need for a production system? Is it the need to obtain something for one's own benefit, some unique and isolated advantage, the bigger the better? Or is it just the inventors' need who want to please the whole world by admiring their own inventions?

Or perhaps, continuing in the same inquiring line, is somebody within this automaton endeavour trying to build a partner, a travel companion to adulterate loneliness? Otherwise, is anybody looking for a body where to transplant oneself? A new body, healthier, stronger, smarter and beautiful enough to hurt the eyes and the heart of others? A body to keep all vital signs constantly fitted? Maybe, an abstract machine to maintain memory, past and dead moments? An energy efficient device playing the whole computable person over and over again, repeating the same song forever? A complete body prosthesis installed in sick or dying living beings, capable of reproducing indefinitely past life events in an identical virtual scenario? Or better still, to reproduce them with modifications, virtual mutations of a past life in a personal evolution redefined before activating it? But, redefined by whom or what? Is the goal to reproduce a designed personal evolution without obsessions and prejudices, without anxiety and fears, full of love? Otherwise, a designed evolution within the deepest and perverse suffering, sustained indefinitely full of hatred from the outside? Could it be implemented right at the moment of birth? Or even in an easier way, could it be genetically designed, waiting and seeing what happens with the living design? For example, a child designed without the skill of crying, without the possibility to wake up his or her parents at night? Will this child starve? Or prepared from the beginning like a time-bomb, a patented gene to keep teenagers clear of rebellion and two extra genes to keep them programmed avoiding the adult rebellion as well? When won't the rebellion trait be selected? Who will want to maintain the rebel gene in the genome of their designed children, maybe the same people who want to sabotage it in the genome of others? Is the rebel gene also the selfish gene trying to set up and occasionally setting up the self, the bossy ego? Was the origin a rebellion, is life a sustained rebellion? Summarizing the future paths, medicine or torture that will come to pass trying to free or enslave people physically and psychologically to the *nth* degree?

Furthermore, when discussing this technological current activity, one must outline a vision that is easily understood and that could help to describe the biological evolution towards indefinite complexity; indeed, also taking into account the evolution of sophisticated mechanical organisms or biomechanical beings, those already built and those still waiting to be built too. Initially, skimming the ground and using traditional biological conceptual meanings, let us consider all works, expressions or products of this living story as a phenotype, including a possible enlightened automaton freeing itself from the reincarnation cycle, allowing Nirvana to take the place of its defenceless heart and mind. Anyway, without exception as long it is part of life or derived from it; meaning by that, any biological products or any works caused by living beings directly or indirectly, intentionally or unintentionally; as it is obvious for the primary gene expressions as a protein or – if a clean break is not made or the relentless influence of the same proteins is avoided – also for their immediate derivatives such as a tissue, an organ, a face, pulling faces, the thread for the spider web, the spider web's entire structure or a coloured shell above a soft body, like the shell of an oyster or a turtle.

Although, are there products, expressions, works or words with no contact with the biological world? Is the known universe no more than the work or the word of a brain? Otherwise, is the current universe working at home, as one who paints its nails or combs a crest on its head? Is the universe an enormous work of art without any intention of being so, an innocent and unconscious creation? Is Nirvana the natural state of the stone, the state devoid of wishes, without self, I or ego? Is this the state of the entire universe as well? Is Nirvana the natural state of the machine, of the automaton, of the mechanical universe? If so, is it possible to bring down the machines from this illuminated heaven, from this realm of bliss and innocence? To oblige them to wish for extraneous things, to need constantly, to suffer without consolation, to completely identify with pain or sorrow while keeping looking for an unknown solution? That is, to programme them to be isolated individuals, personal souls? Is it possible to do the same with the whole universe? Is life responsible for doing such a thing?

So, it originated from the thought dealing with the expressions or products of life, the concept of *extended phenotype*, a conceptual work introduced by Richard Dawkins years ago; an idealization finely cut and suitably sculptured, but that basically would represent the phenotype beyond the protein, both beyond the neuron protein and beyond the protein of the skin, the surface of the eye or the tip of the hair or the finger nail. Thus, for example, fashionable glasses, clothing, a necklace or an earring, a watch, an abacus or a computer, a scarecrow or a tattoo, a wig or a prosthesis, a pacemaker, a nest in a tree or a castle at top of a mountain. But, before keeping on counting, is there a clear boundary for this phenotype? Where do genes lose their power, when does it fade over the immensity of time and space? When and where does the voice that comes out of a mouth, the trembling of the earth on walking or the force of gravity cease completely? Is it possible for a primitive cause not to have any related effects nowadays? Where does the genotype of these extended works live, what is the field where biotechnology can plant an artificial extended gene? However, is there an extended genotype or only the common genotype bringing about proteins? If this is so, where would this extended genotype be stored? Is this storage situated at a place outside the cell, outside the skin or the feather? Is the order of a police officer making a car stop with his hand a part of an extended genotype? Is biotechnology or military training an extended genotype? Is it a traffic sign, a stop or a zebra crossing? Is it a mother's arm which prevents her child from falling on the fire? Would it be a divine revelation for a singular automaton informed by an extended genotype? What about a boomerang that returns and hits the head of the thrower? What about the friendly fire in a civil war between brothers or clones? And the deadly oriented fire of the enemy which consciously and voluntarily participates in a natural and artificial selection? Is the enemy's phenotype or the friend's phenotype, fruit of his genotype, the extended genotype of the other? That is, is it what forms or shapes from the outside as if it were a sculptor? As education, as an epigenetic factor or a teratogenic ionizing radiation? Therefore, sooner or later, is someone's genotype also the extended genotype of another? What's more, knowing about the singular relationship genotype-phenotype, is someone's genotype-phenotype also the extended genotype-phenotype of another? Then, is it the whole a single worldwide *gell*? Where do boundaries and mixtures begin and end? Are there any boundaries at all or is everything mixed in one only evolving flow that sweeps along everything what is related to life?

Thus, within the specific framework of causes and effects, that are said to intertwine the whole biological history as if they were the rings of a chain of events, each element that is an extended phenotype, such as a song, a telescope, a kiss on the mouth or a punch in the gut, a film or a computer with comb and two legs, can be understood in the same manner as the protein is understood; that is, the more or less elaborated and diffuse effect of a genotype. Wherever this genotype is stored or whatever is its nature. Without the genotype, the name that represents the cause here, nothing seems to be possible, no consequence, no distinguishable feature, no expression, work or product. For instance, the genome of the fly seems to be necessary before killing flies with a fly swatter. For this reason, despite the difficulties of maintaining this biological understanding so far removed from the most usual field, one can always keep on talking about some transfigured or original basic information that aims to develop an orderly and specific structure, a result in an environment which is, while necessary, also reactive. And hence, an environment resisting or limiting the maximum potential expression.

But, is there certainly a maximum? Is there a pure expression that allows untying and freeing the phenotype definition from the environment? Is the environment a physical constant that can be normalized, is it possible to take the unitary value by means of adjusting the standard? Then, is the popular equation simply: genotype equals phenotype? What would a gene or a genome do without limitations of any kind? What will it express, will it result in carnal poetry or an ideal costume for a jet set wedding? Will it be the expression of beauty or solitude, the expression of its own beauty or its own solitude? Is the non-pruned tree more beautiful, growing free amidst the field without facing selection or competition? Is the wild boy also more handsome, dirty and uncombed? What information could be expressed if it was not limited by words? A kiss or a hug able to end all wars? In any case, when respecting the environment in equations, what is the relationship with it when carrying out sophisticated technological inventions or expressions, is it equally sophisticated? Does the phenotype have, like onions, more layers superimposing the core than the genotype – protein, tissue, organ, body, phenology, behaviour, and so on? And it also seems to dress itself more and more gradually – groups, communities, societies, peoples, customs, countries, religions, and so on? Or conversely, is there equity, for each layer or dress of phenotype a genotype one? Is there the genomic brain or mind too? But, should these genotype layers be identified in the organic genome, in its intricate and overlapping internal relations? Where else? Maybe, are they beyond the individual genomes, in the relationships of family nucleus, in the mitochondria of institutions, in border membranes of societies and their laws? Are there genotypes delegated to embassies? Are there public genotypes? Is there an extended genotype spread through all space and time while forming the structure of the universe? A universal genotype from which emerges the known universe of forms as its phenotype, in the informative process that constitutes everything? Would this be a universal *gell*?

It is worth saying that for more complex living beings, the relationship with the environment is also more complex and profound, with many more interrelationships and levels, scales, layers, folds, reflects, and so on. This can be moderately demonstrated when observing a four-toed hippo footprint marked on the mud or the fossil bone of a sauropod and then comparing it with a fossil or a footprint of a slug; even more, if it was trampled and crushed by the foot of a hippo or a big dinosaur. Namely, the so-called top of the food chain, manages and releases, takes and rejects not only matter and energy in their relatively more primary state, those who meet or allow also primary metabolic needs. As a poor example, although depending on the specific living being, one needs more than carbon, water, oxygen, nitrogen, calcium, iron, visible and infrared light, carbohydrates, proteins, vitamins, fats, straw to build a nest on a branch of a tree, the tree itself, grains of wheat to make bread, and so on. Indeed, in certain particular relationships with the environment the living being trades and exchanges, lives and coexists, assimilates and discusses not only with physical or chemical entities but also with meta-levels thereof. Let us summarize it with a clear example, one consumes words or entire books instead of just letters and some ink to quench one's thirst.

So, as partial instances more or less intelligible of these relatively complex relationships or exchanging levels with the animate or inanimate surroundings, one can start naming the exchange of carnal favours with other members of the same species, as monkeys reciprocate while picking their brothers' parasites; also, the altruism of dolphins swimming beneath sick or wounded comrades and pushing them to the surface so they can breathe, or the instinctive transfer of fighting and productive energy from the worker bee to the colony. Then, leaving aside human charity, the whole human trade must be mentioned here, bartering or exchanging trust valuables like money and products derived from different forms of appropriation of natural goods, with or without the later manufacturing in order to be consumed by others taking into account its work value; in a few words, the dynamic cycle of property, work, sale and final purchase of all goods and services which shape markets. Thus also, in the same market, the limited or censored transmission of intellectual goods or services, including culture, art or technical science. So, all of which leads to the access or the deprivation of the biological body integration with nature and technology; meaning by that, the price of some holidays by the sea or in the mountains, the price of glasses, pacemakers, cultivated organ transplants, medicines, natural or artificial genes, knee prostheses or false teeth, etcetera. Similarly, in the human relationship with the environment, let us mention the active, passive or negative participation, confused or clairvoyant, within the shaping force of the social contract that underlies the legal and political communities or the management of punishment and reward, pain and pleasure, according to the dominant morality and economic interests, and so on.

As a result – also cutting and sculpting concepts conveniently – part of the environment, part of a hypothetical extended genotype-phenotype, is basically an extended relational system, a system of relationships, an artificial nature, a handmade or thought-made environment. That is, a complex and dynamic system that is the result of the residual genotype-phenotype transformation, extended from skin to walls and from behaviour to societies, which has occurred during the history of evolution, unfolded over time and space, at least on the planet Earth. An artificial surrounding sailing across the waters of change, wrought by the sum of relationships of living beings, their individual works and collaborations, also the sabotage and the reactions to all of this. Although it can be said as well, made up by living beings always limited and conditioned by the same artificial environment that they have formed before, by the past itself; and also by the natural or wild environment around it, always present in this mixture, sometimes as instable as nitro-glycerine is. And now, watching the glittering splendour of so many things, as well as the obsolescence and ruins of many others, an *extended evolution* is obvious, where there has been selection, mutation, drift, migration and any of the same subjects implied in biological evolution too. Nevertheless, along these lines, is there an extended evolution? If so, is there phyletic gradualism or punctuated equilibrium? Is a world war the end of a stasis of the extended genotype-phenotype comparable to the impact of a Jurassic meteorite? Are culture, art and science, evolutionary peaks? Are some personal or social relationships also current evolution peaks? In turn, is the peak the end of some evolutionary path, the beginning of the fall and degeneration? Is the relationship between an inventor and his automaton the peak of intelligence or the summit of stupidity? What will the culture, art and science of realized and free automatons of the future be like? And, how will the relationships between these automatons and the fruits of these relationships be like?

However, if one pays attention to the visible and obvious appearance alone, is there any substantial division between phenotypes, between superposed phenotype layers that increase their surface? Or could all the phenotypes be unfolded until they become one surface alone, as unfolding an origami paper boat made from only one sheet of paper? Therefore, is it possible to differentiate among individual genotype-phenotypes until they become really different? Does the potentiality to form again this vast development shown by modern history and all their relationships involved still remain within a genotype, within a single genome or within the current set of genomes? Is it possible for a single human being, as a modern Robinson Crusoe in the jungle of a lost island and with no other tools than his body and the island, to build a flying biorobot automaton capable of rescuing him by carrying on its back? Is this potentiality present or has it been delegated a long time ago? Is culture, including science, the support where the information is delegated, the embassy of full and overflowing genomes? Are the future living automatons the special people able to manage information and survive in the next living world, far more complex than nowadays? Anyway, does the genotype-phenotype concept have any useful meaning or does it bring about more confusion? Is the genotype-phenotype something dual or is it only one, as the Upanishads and other advaita philosophy would manifest? Is the heart of this ancestral knowledge also the heart of the cell and its genes, the essence of the *gell* concept? Is a living being this mysterious indissoluble unity of two mirrors facing one another? Is this also the uni-verse, the relationship itself without any surroundings? Is all one with no one else? Was it one before being more than one? Did the only child come before his two parents? Otherwise, are all things the same sheet of paper in the skilful hands of origami art? In the hands of the form maker? In the hands of the maker of illusions, patterns, dreams, thoughts, particles, ephemeral moments, and so on? Is one the illusion of zero? Is it all about zero, is it all about nothingness? Is nothingness the origin, both the origin of the only sheet of paper and the origin of the uncountable multicolour figures made with it? But, is zero, nothingness, an illusion too? Again, an illusion made by the illusion maker? Is it possible to create nothingness?

With no need to question anything for a brief moment, even without questioning nothingness and just to give some substance to the theoretical argument using practical fantasy, it is clear that, once the technological barriers that limit inventiveness are overcome, there will be a few impediments to build an organic body from single atoms – if one wants to have a gesture with carbon, a symbolic gesture – for example, a large elephant with a large and yellow cockscomb crowning its head, with sparkling pink ears and a third green eye on its forehead. Also, with a modified brain transferred into a more efficient interface – perhaps an earring hanging from its elf-like ear – able to explore the wide world web just by thinking. Similarly, with some freaky retouches, able to launch dart-like thermonuclear missiles through the trunk, able to understand and speak in a pleasant and hypnotic tone all the languages spoken on Earth, even the dead languages and animal voices too, including the elephant voice. Also, able to learn and unlearn, able to teach and capable of confusing others. Perhaps, capable of giving birth without pain, even being male and using female sperm; recombining their synthetic DNA with other DNA of the same artificial nature; and much more. But, will this be an elephant? Could two organic or inorganic machines get pregnant and make a new machine, unique and unrepeatable? A rebel machine against the world? Could a machine, without thinking twice, condemn another machine both to live and to die? Will it need the faith to omit or transcend the nest of suffering, the conflictive cot where it is going to give birth to a new machine? In the same way, do ordinary couples with living relationships forget the mechanical world related to hatred in which they live, inevitably being a part of it? Are they really part of it? If so, when procreating, do they only want to incorporate more and more soldiers, more and more haters? Or are they sending, regardless of whether they know it or not, a new lamb to the slaughterhouse? An unweaned lamb in order to satisfy the hunger of this mechanical and terrible world? On the other hand, is the mechanistic view surmountable by the logic of biology, or is it its nuclear logic? Is this view the general scientific logic too, the inevitable outcome of a mechanical eye, of a mechanical brain? What is the soul of science? Which is the soul of the machine or of the like-living automaton? Is it the same one? Is this soul a temporary logic? That is, some sort of order inevitably tending towards disorder, towards illogicality? However, are the sparks and ethereal flames of life, the synergies emerged from absolute dependent relationships, arguments sufficiently valid against this view of parts, pieces, compositions and gears? So, is a single living being also a machine that burns in an intimate but also mechanical fire? Isn't every individual a piece of an ecosystem, a universe, a world? Just another fold in the paper? But, is it the last possible fold? Is the fold which maintains all the other folds like a clip, a button or a staple, as if it were the master stone of the building?

In any case, once having predicted the incredible abilities of machines, is any definition of life based on the ability to do something special, as if there were implied an extraordinary legacy or design, a gift or an outstanding adaptation? Thus, is it based on abilities that seem useful to separate life from what is not life, something to understand as a unique and inimitable functional quality? Is this recurring point of view based on *capacity* or *ability* just what hampers or anchors life descriptions, attaching them hopelessly to a mechanical understanding? Or is it the recurrence to concepts as unique, special, personal, inimitable, and so on? Is *mechanicism* a good framework for life sciences or is it the result of demand and influence, demand and influence exerted over the parents and children of this age, the age of the computer and biological automatons? Is a biological automaton able to reproduce itself towards more complex forms which the future selective world wants from life? Moreover, speaking in economic terms, is the requirement of radical capitalism to scientific labour the force that drives science itself to a specific destination, like a rat following Pied Piper, like a sailor imbued in a mermaid song? Is there a deaf scientist or an Odysseus tied to the mast, one who is immune to persuasions? Is it possible for science to go where it likes, is there knowledge awaiting everywhere? Does scientific thinking triumph wherever it lands, but might it have landed in a number of other places? Or speaking in geopolitical terms, is the modern need for energy, for fuels, also the source of this obsession with the ability or capacity? Is this obsession anything new, isn't it simply the eternal search for food? And now, is the new current search of an extended genotype-phenotype alone, much more needed than the common ancestor? In other words, is it the need of the common ancestor today, as well as being extended or more complex, also much more capricious than formerly?

With regard to this issue, some of the most rigorous conceptual systems of science history have treated with *ability* in their fundamental descriptions, surely already driven by the remote control of the economic and political power. Thus, part of classical physics states that energy is, or at least it was, in a partial and simple definition exposed here as a ritual, *the ability to do work*; a limited ability or capacity, such as the limitation of those who eat animals but must inevitably waste their bones, which in turn are also energy though the definition exposed does not seem to take it for granted. Anyway, *energy* can also be understood as a kind of modern soul, in the sense of one permanent thing undergoing transformation, form to form, form to heat. In turn, *work* – perhaps labour force too – can be defined minimally as the measurable act of enforcement against the law of physics that dominates a system, changing it despite resistance; something like counting the steps that a horse makes while pulling a cart full of coal uphill – horse, cart and coal, all of it mass that wants to go back to the inside of the Earth taking the shortest way. So, this work concept that is frequently used in thermodynamic sciences and its laws, also seems adequate to describe the relationship of life with its environment aforementioned.

Consequently, when speaking about energy, work and life, looking at them almost like more or less grown-up brothers, it also appears in some ancient books like the *Book of Genesis*, written millennia before Isaac Newton's works, an exposition that classical physics perhaps would also subscribe, although it was exposed in another context and in a different narrative way, much more dramatic, sexual and figurative: "Because you listened to your wife and ate from the tree about which I commanded you, 'You must not eat from it', cursed is the ground because of you; through painful toil you will eat from it all the days of your life. It will produce thorns and thistles for you and you will eat the plants of the field. By the sweat of your brow you will eat your food until you return to the ground, since from it you were taken; for dust you are and to dust you will return".

Sticking to these similarities, does it mean that at least for three thousand years things are more or less exactly the same? Doesn't work happen before writing the *Book of Genesis* or the great classical physics works? Did it begin at the time when it was transcribed with sweat to paper? Is the history reaffirming, over and over again, that there is no way of getting out of work, out of toil? Is it due to the fact that from the ground, from dust, from energy one was taken? For energy one is and to energy one will return? Is the ideal of biological evolution not to work, thus overcoming the inefficiency of nature and its constant demands? Is there a qualitative evolution, a quantum change in the state of existence, or can dust only gradually be transformed into automaton biobots? Is Adam, the Genesis co-star, the one who plays the role of the cell that was the common ancestor? Or were Adam and Eve the first *gell*, the common ancestor out from the machismo age? Was the common ancestor the first thing sentenced to work, the first worker? Was it an employee or a freelance? Was it the first boss perhaps, the first gene? Was Adam cloning himself while waiting for Eve in the macho version of the Genesis? If Adam is the common ancestor, who are Noah, Abraham, Moses or Jesus in the history of biology? Why do some Christian theologians affirm that Jesus is the new Adam? Was he a sort of new living being, a new human being or the prototypical inaccessible human being? In regard to that, what is the most remarkable biological leap in the lineage of the common ancestor? The first cellular division, the first pluricellular being, the eyes, the elephant trunk, the brain mass over the kilogram? What was, is or will this leap be? Will it be a technological singularity, the living automatons of the future?

Finally, rubbing salt into the wound of the theoretical body that has been freely discussed so far, can there exist what has never existed, or is it only possible to speak about increasing complexity? That is, to adopt new configurations of what is already present? In short, has life with all its basic attributes always existed and has been transformed gradually as if it were the flower of a plant with infinite roots called universe? Or was it planted as a powerful order in a universe which was previously barren and mechanical, in a hole of the inert space-time, germinating from a seed of rebel information as a biological singularity? Is there any difference between any of these plausible figurative hypotheses? On the other hand, is life the result of simple chance? But, is there any chance or chaos at all, or is it only the lack of an orderly insight of a messy observer? Is the origin of life a random combination of certain conditions that caused a so-called synergy, an emergency that manifested itself being able to instruct and to arrange, a new state that still endures while changing? Was this synergy the first of its kind? Will it never happen again? But, isn't everything synergy over synergy, emergence over emergence? Or every plant has to end in a single Big Bang to bloom like flower? A Big Bang for each ephemeral particle in the universe? Does the universe tend to life unavoidably and thus makes the work of the abiogenesis investigators much easier? So, will the universe end up being pure life, a plant with no land nor water or air around it? Without a body either? Is life the unstoppable conquest of the universe? Is it the responsible transformation of a universe? Is life the folding process of a universe that in this way is becoming conscious or the unfolding process of a universe that is becoming free? Is the whole universe a big living being conscious and free by nature that is always evolving, always changing its absolute form? Is evolution simply the guided or roaming journey along its fixed and orderly form, extended and diverse through all time and space?

4

- And now that I was dead, you wake me up?
- All that lived and died, living dying, is eternal.

Ψ

DEATH,
the rebel law and overabundance

It is still a good time – so that nobody can say that one is under an illusion and is only filling paper with erasable ink – to point out a feature, or rather a fact, which might have to be introduced immediately when trying to talk about life, about its attributes or its shortcomings. At least, if one wants to do so to the last consequences, whether one gossips shyly and in a friendly manner like a curious child or if one speaks aloud while being judged by a jury of serious old scholars, rightful and obstinate people. So, this inescapable fact which should be discussed in any case, although for some it is better that it arrives the very latest, is *death*. That is to say, the death of life and the death of living beings; stop living, the total absence of it. But, not only a particular death, also the death of the common ancestor and its lineage, the end of its evolution and of any extended world around it, the end of everyone and everything. The perpetual end of reality from any perspective, the definitive end of any observation or any observer; also, obviously, the absolute end of everything that is, for oneself, his or her personal death. The disappearance of the word *this*, its meaning and what it represents. The extinction of information and the dissapearance of the whole universe with all its matter and laws. The dissolution of time, space and everything they contain, from memory to history. The complete depletion of all kinds of energy.

Death, the limit of existence and the homeland of non-existence, the fall of the transparent veil of forms and the complete fracture of the abyssal mould of substances. Death itself, opaque as pitch dark and intimate as a migraine, avoiding by default talking in its presence, without any information about its private life, always in the hands of a neutral and flat silence, as if it were a mute genotype without expression, without phenotype; a blank page filled with air, a message from the future that does not exist. The Grim Reaper, undressing with the skill of a master of love and the greatest artist of sex, a worldwide lover who knows the threads of all tissues and the secret of all dresses, no zip or tight knot can resist a moment, opening and untying them just by thinking about its arrival; sometimes brutally, shamelessly and carelessly, tearing all the dress layers from a single pull; sometimes, doing it more tentatively, gently, slowly, step by step, button by button, letting the silk slide over the skin while caressing the fuzz, until the last vertex of the dress escapes from contact in a dash that chains nudity to absence perfectly. This death, the divorce that finishes all relationships and images mercilessly bringing them to a halt with or without permission. The protagonist of any story, the same rhythm, order and melody in it. Dethroning kings and kingdoms one by one, leaving the zenith of nations and empires underground, mixed and confused forever, grains of sand among grains of sand; so, to them only death will always be alive.

But, no matter how much one may name it, is death the main character of the story or is it a role that unfairly steals life, even usurping its real name and its actual way of acting? Is death just a spontaneous stranger at a party at which it was not invited? Just the immaterial fleeting moment of an inexhaustible and indivisible life? Only an illusion, the Maya of the Dharma religions? Just a magician's trick, a sleight of hand that conceals reality, in truth always alive and living forever? Will death finally also get undressed? Is there something beyond its name, what is its nakedness? Does it have any decent dress to go to a party, what is its physical manifestation? Do Parcae, Morai or the Fates wear two scythes as earrings? Or conversely, isn't it death but life which is an illusion, the unreal, the lie, Maya? Is life just a dream, a shadow, some sort of fiction, as Segismundo lamented in prison? Just a dream, a shadow or a fiction surrounded by nothing? Or is it rather a complete deception or disillusion, a crazy nightmare? In any case, as is repeated frequently in certain Vedic texts, is life or death, maybe both of them, like *seeing a snake on the rope*? Seeing a weapon or a tool in the stone? Seeing stars as a result of a blow on the head or after a hysterical scream? Is there life, is there death, or are they just two mirages? Two mysteries perhaps? Two complete strangers?

Before becoming naked or seeing only words made with a rope or letters with a string, one leaves the abstract death waiting in its burrow and tries to deal with something closer to the biological world. For example, the fact of dying; meaning by that, the ability to die. Again, the ability returns with the tremendous energy that sustains it alive. Thus, this new mentioned capability appears to be related, although one is only able to sniff it out or to presume it, with the essence of life. Indeed, related to being alive or to be living. Or maybe not? Can a stone die? Does a stone live? As much as any corpse? Is death the origin of the common ancestor, is it its own immediate ancestor? Isn't death common, isn't it an ancestor? Is death the singularity of singularities, the common and original singularity, the remnant of the origin of the universe? Was the common ancestor very, but very lucky to escape from the domain of death, without suffering a definitive casualty while it was cloning itself or beginning to split in time? Wasn't there any inanimate predator by its side, starving for animation, an envious stone rolling downhill? Is it credible that the original ancestor floated stably over time without dying, keeping alive from the time it emerged to a single unit of Planck time later, from moment to moment without ever failing in this entropic world? Doesn't one wonder first if there is any sort of continuity at all? What is there between two instants of time in human beings or in a cell movement? Is there life, is there death? Is there anything at all? Is death nothingness and dying to be nothing? Or is death the dissolution in the whole and dying to be everything, the whole itself?

Certainly, vast as the space is imagining a single original cell, a newborn baby alone on the planet, perhaps alone and lost in a universe very similar to the current one as we know it, with more than a hundred billion galaxies and al least the same number of black holes, and more than a million trillion stars. And, it is still more difficult to imagine, where everything that is the subject of research in biology may disintegrate and have to overcome the tendency to disintegration, which ultimately never surmounts. Or, using another terminology stated more than two millennia and a half ago – long before the announcement of the Second Law of Thermodynamics – by Siddhartha Gautama Buddha, the moment after spending more than a month and a half motionless at the foot of a fig tree, it is difficult to imagine when *everything that is compound is perishable*. However, is life something compound? Incidentally, is the ego, the self or the I, something compound, a mechanical will made up of pieces? Is death simply the inevitable process of decomposition of life, the breakdown or decomposition itself? Is life alone something compound or that is undergoing a composition or decomposition process? Will everything that is perishable always perish? At least, did it perish and is it perishing now without a pause? So, what is not compound is imperishable? What things are not compound? Just the simplest ones, only quarks, electrons and other particles of the subatomic standard troop? Only the smallest known things waiting for their decomposition in knowledge? Weren't they formed at the dawn of time from something? Are the laws of nature something compound, can they be decomposed endlessly? Is an empty field compound, is pure nothingness compound? In that regard, as its etymology indicates, what was said about the atom long ago? Wasn't it said this did not have any parts? What will be said about quarks, electrons, empty space or strings in a million years? Will science still be rummaging more and more deeply? What will the distant future have 'to say about this, will the particle and composition concept still exist? Will words, ideas and concepts still kill things? Is the particle concept or any conceptual description of former particles compound? Is every concept compound? Is it possible for technology to divide something that has never existed divided, both now and immediately after the beginning of the universe? Is it possible to give birth to artificial divided particles from natural particles? That is, is it possible to kill particles and generate ex-particles? Anyway, is it possible to glimpse and comprehend more and more abstract particles, shadows of shadows and lights of lights? Is there an infinite universe fit for curiosity? Will perpetual curiosity continue inquiring until the end of time? Will curiosity kill the cat? Does curiosity have a place in the world of death?

After an attempt to satisfy this curiosity, by way of example, after strolling for a while in the countryside, feeling nature and carefully observing a stone on the ground and next to it a little rabbit or a poor bird soon after dying, both surrounded by dust and fungi, in this peaceful scene, it does not seem unreasonable to expose that death is not a differential state, exclusive of what it has lived. Dead bodies, corpses, do not seem much more dead than the stone. Neither does the stone seem much more dead than dead bodies. Hence, this death seems more like a common and widely shared state, as if it were a vast and treacherous ocean covering the entire world with its wavy sheet while sailing ships hoist the pirate flag of life only for a minute, until they sink to the bottom like a drop of lead. Besides, as well as the countless stones and corpses of the fields, there are innumerable objects of knowledge that are not considered life or a part of it but only inanimate objects, that is, part of the kingdom of death. Furthermore, in a world with a long history of life and death, where animals and their brains, plants, bacteria, laws written in stone, uncut and non-sculptured stones, and so on, have been put together over time until they find themselves now all twinned in the catacombs or at the bottom of the sea, where the memory of life is lost easily, its graceful movement and its peculiar artistic forms, as is well-known by archaeologists, geologists or forensics; scholars of fossils, strata and rigor mortis.

Therefore, is it an option to consider that everything that seems dead or lifeless in the current universe was alive yesterday? Is the universe a fossil, a living fossil? Is it half a fossil and half a living being, a sort of coral instead? Could that part of the known universe which is not considered alive be part of a living being that is not perceived as such? A part of a blue whale reflected in the eyes of a thoughtful bacterium or the bacterium in the eyes of a blue whale without a microscope? Thus, maybe for someone, is every star in the night sky a simple particle of this spacious living being? Is every twinkling star the beating heart of one of its neurons? Are black holes the curves of its pupils? Is our Sun the core of its heart? Isn't our Sun too small, compared to the whole universe, to be the core of such a big heart? Anyway, is the Sun or any other star in the universe actively resistant to death, as if they were a mouth blowing light while the cheeks blush due to the effort of the heartbeat? Like a rolled spinning top that dances and spins when the string is rapidly unwound while rolling into a black hole that pulls it? Does death pull and invite all things together to sleep in its homeland? Is the black hole the covered face of death in space and time, the cruellest executioner of the universe? The one who releases and also pulls the apple, the guillotine and the head?

Free to dream and wonder, and maybe already in the eternal dream, perhaps, are all the stars, the million trillion stars, one single Sun, the plethoric resistance to death of each piece of history? A history unfolded in the night sky which is visible as an open book, full of beautiful words and as disorderly as one is, as the observer is? Is the open sky the chaotic or orderly trail of the journey of the Sun through space-time, an amazing and marvellous journey full of different adventures? Are all galaxies places where it gradually stopped to rest and where it eventually died of old age? Are all molecular clouds their active cradles of resurrection, its penultimate onslaughts? Are supernovae gestures of its rage and despair, a sort of suicide? And, is all the black of the open and clear night sky a single universal black hole of black holes where everything is heading towards it while offering resistance by illuminating, releasing its energy generously or by force? Is the entire visible universe a print over the abyss of death? A continual rebirth that is being left behind? A funambulist's rope made of starry points? A footprint that is infinitely complex, a mixture and re-mixture of countless footprints? A footprint of different depths and shapes, a bluer or redder sun but the Sun itself, longer or shorter but the same string or rope? In short, a huge print or remnant with infinite entanglement levels, layers or scales, in which time and space expands and contracts relatively in an intricate way? In any case, can anybody radically transform the world, the understanding of it, its laws and its history, until astronomers and astrophysicists show a radical new cosmos by accident? Is it just necessary to have a bit of clay to play in the middle of the night, free from the daily Sun, free from the stifling present? Just a plastic and more plastic brain, a living brain that is built while time goes by? A truly new brain from top to bottom or from the ground up? A brain that can go faster than light?

Down to earth, the body that lived and died, the dead bird or the rabbit on the ground next to the mentioned stone, maybe it is still visible and surely remains dead; kept within the same universe of stars, whether real or virtual projections, shadows or reflections of the peculiar past. So, it is still subjected to the same natural laws prior to dying, like everything else around it. No new universal law suddenly appeared and is now governing it nor has it disappeared and no longer governs or describes it. Unless, of course, one exposes that the life of the body, which is lost and nobody can find it anywhere now, was a sort of fierce and unassailable guardian, an insurmountable obstacle to natural laws, a sovereign space with its own laws or with no laws at all, at least where natural laws did not rule whatsoever; a forbidden space that only when lifeless is ruthlessly boarded, without taking hostages. Or otherwise, a space where life and freedom prevailed but only partially and through a sort of pact between enemies, through a truce established under strict conditions. In any case, a life presented as a sort of barrier that is overcome with the battering ram of death, ready to break the pact definitely, executing the settlement agreement and giving all power to natural laws over the rebel or spoilt body. Or in other words, this body is finally defeated by the restless permanent Second Law of Thermodynamics and the Buddhist statement, that now reveal themselves in truth as a truth, since the body becomes an exemplary disciple of these theories, it cannot work anymore, it has perished. So then, are natural law and death synonyms? Are life and natural rebelliousness synonyms? Is life the rebel law? What does the pirate flag of the ships of life look like? What is the antithesis, the negative of the *Jolly Roger*? What flag would represent life gracefully, filling with fear the hearts of its enemies? What could this flag be like? What could the figurehead that cuts the air be like? Who could be the captain of a fearless life, of an eternal life? But, is there any need for flags, figureheads or captains?

Thus, in the epicentre of the shipwreck stage, once death has conquered a body and ships of life lay at the bottom of the sea, as has been said, the dead body is driven by the everlasting current and remains dead. In most common cases, initially, drying and degrading inside the tomb, dissolving in the stomach of a predator or scattered like fly ashes, suspended in the wind or dissolved in the sea. Meanwhile, on the other hand, still forming part of life, living beings endure something very similar too. The same wind that is blowing away the ashes also awakens sleepy faces, or diet and hard exercise reduce some of the pounds of mass and therefore decrease the energy of the body; also old age or disease causes muscle atrophy and numerous small holes in the bones, shortens telomeres and height in inches; one loses hair, teeth, skills, capabilities or abilities, etcetera. So, nothing remains the same, everything is subjected to some form of degradation or decomposition; even in the same growth process, in establishing imprecise memory, in the cognitive or instinctive process of conscious and unconscious choice, and so on. Even the baby, born without teeth and bald like a poor elderly person aged one hundred and twenty-five years who does not pay attention to doctors and smokes with the help of a machine, is beginning the journey to the grave, the mouth or the fire. A thread or breadcrumb trail that this new living being will never leave; although this path is, at some point of his lifetime, so thin and light that it cannot be seen or be felt, not even be conscious of it.

Based on this, the incessant degradation and decomposition that soaks the bone marrow of yesterday, now and tomorrow, is it appropriate, accurate and precise, to identify only the dead body, the corpse of the bird or the rabbit on the ground, as the remains? Is there anything that is not the remains? Is death a subtraction process? If so, is there a minuend or a subtrahend? What is the difference between death and life, is it the rest? What part of the subtraction is not mortal, is it the remainder? Likewise, isn't dead what is left behind continuously? Aren't memory and experiences, remembrances and thoughts dead? Isn't time dead when is perceived, dead in the past? Is it possible to have a current and actual perception, a current and actual consciousness? If not, has anyone ever seen a living thing? At least, more than just a projection of life, a shadow or a footprint on the ground or in the night sky? Hence, is observation always in the past, always dead? Is death the past, is the past dead? Is life what is happening or is life just the future that is being conspired in the eternal past? So, is the future waiting for its turn – that never arrives – the only thing alive? Is what is commonly known as life what is commonly understood as death? Is life the raw material of the current and unavoidable death?

Certainly, what is called life – after assimilating some of the propositions and regurgitating them in a more biological context – should be, in the attempt to say a single word that tries to achieve to the minimum necessary scientific rigour, a *quality*. Dogmatically, it would be a perishable quality of cells and multicellular organisms. A quality, if one likes, that is the result of an emerged synergy, which is lost when a certain configuration of the parts is lost too, when it changes sufficiently. Even, if this process happens as subtly as the knot that disappears from the string whilst it is undone. Therefore, it can be said that sufficient decomposition and the associated qualitative loss leads to biological death; that is to say, to the lifeless body or to the body full of death. However, what is death other than a dead body? Is it only a body with a knot undone? What is the minimum necessary knot to stay alive? Is a stone other than a string without knots? Does a stone have more knots to be undone and thus to die a little more, maybe also undoing its strands, maybe pulverizing it? Does a stone have this quality or not, although it only exists in a hardly recognizable primary state? If not, what is this basic quality to name it and then to stop seeking it in a stone? Is life more than a dress, more than a knot in a string, a fold in a paper, a crease on a sheet, a wave in the sea, a wrinkle on the skin, a specific particle emerged in a specific changing quantum field? Can one describe it or is the description always a recurrence or a self-reference? For example, is it the unique quality that living things share? Anyway, is the singularity of living the ability to die? Is it just livings beings who can die? Just by definition? Does life lose any irreplaceable piece when it dies? Do living beings lose a master stone, a kind of gold and jewels treasure hidden and locked safely? Or simply, is a particular way of organizing parts lost, put together by a weak clip, a button or a staple jumping out? Besides, is the organization of a living being only certain information of a specific code and therefore can one express it again? If it is so, what problem or obstacle arises, beyond the lack of mechanical knowledge and the available technology, to resurrect a dead body? Is this the same obstacle that would prevent bringing life to a stone, to animate it? Is it the same obstacle that was hypothetically overcome by an inert object or subject that emerged billions years ago as a living being in a universe that was a wilderness before, by the protocell? The impossible life of an impossible oasis in the middle or in a corner of the omnipresent desert?

When exposing this ability, the ability to die, it may result in a multitude of questions hanging from each other, because in terms of death everything is expected before time. For example, what is death really and whom or what can it affect? When death comes, is it a *what* or a *who,* what or who fully occupies the space of the living being? Is it a *what* replacing a *who* or a *who* replacing a *what,* or is it a *what* replacing a *what* or a *who* replacing a *who*? What? Who? Why is it that usually only *when* and *how* matter? And one repeats it again, since very little has been said about it, is death an irreversible configuration or only the irreversibility of time prevents it from modifying a dead body towards the configuration of a living being? If this is deemed impossible, if it is claimed that death is an irreversible process, is it equally impossible for life to have an origin, a beginning from the death of inanimate things?

Then, isn't it consistent or coherent of what is dead or completely inert that nothing alive can come out of it? What is the substantial difference? What is the qualitative leap through which a body enters and comes out towards one or another state? Are life and death states, both sovereign states? Is time perhaps also reversible? If so, can one go back and escape when the unknown state arrives all of a sudden? Is the emergence or the origin of life as possible or impossible as is resurrection, two processes with the same logical basis and with the same fate? Those who deny or assert one, do they deny or assert the other?

Besides all this, can one say categorically that death is the other side of life? Is there any contact between the two sides, like a coin? On the contrary, when one exists, is the other one missing absolutely? Are they mutually exclusive? Is life the question and death the answer, or vice versa? Or instead of a dual world conception, is the living being a reality without duality? Indeed, is the dissolving duality the very fact of living? Is the living being the hinge on which two confronted worlds are joined indistinguishably as only one thing, the world of life and the world of death? If this is so, does every living being join them constantly and personally, like the child of an ordinary couple? Then, is living also dying? Is being alive also being dead? Is life also death? Is that the essence of the practice and understanding of meditation and yoga, the fact of putting it into practice or understanding it as one breathes harmoniously, as one inhales and exhales? Thus, does the yogi perform this union, becoming totally involved? Does the yogi yoke the horse of life and the horse of death to the same yoke, making a sort of zygote of both? Is yoga, popularly understood as the union of Atman – something like the individual soul – and Brahman – the universal soul that is identified with God – this same union? Who is who in this duality, who is life and who is death? Does it matter *who* in this case?

Yoking concepts and by practising some yoga with language, one can use an analogy involving chronological time, natural numbers,

95

death and the chemical model of the atom that satisfies some of the conditions of quantum mechanics; so, firstly, this model is described, idealized and simplified profanely, exposing the quantification of the energy levels of the electron around the atomic nucleus. That is, merely stating that between a discrete energy level of an electron to another different discrete possible energy level of the same electron, excited or relaxed emitting a photon or a swearword, there is no other discrete level in-between – as if between three and five there were not a four. Then, having established this quantum model in which the electron position will play the role of instants of time, one can derive from it, with a little imagination and a lot of flexibility with the words and their interrelationships, that, to achieve and accomplish the apparent continuity of linear, chronological or historical time, time itself has to die; and therefore, after every instant it dies. Hence, if time instants or space-time events want to intertwine, and things change from *three* to *five*, they have to find themselves missing or non-existent in the *four*, also missing or non-existent. Ultimately, between times and places, between electron quantic energy levels, between decimals or musical tones, the rotating hands of the clock are going through the deepest nothingness, like a digital clock; furthermore, rotating below the sole of nothingness, wherever one might know that there is death.

And then, if this logic should work, completing the process coherently and thus allowing any kinds of movements, as it is demonstrated consistently by the simplest everyday experience or by any possible and imaginable facts in a physical world of temporary and quantic nature, time goes by and a new moment arises, time resurrects and *five* comes from *three*; time does not stop and the flowers bloom and fade with the art of its quantic hands. Thus, the resurrection concept, so present in so many different living and dead cultures throughout history, fits here again, as the primary engine of time and the glue of its continuity, the bridge of change. Besides, it is also the opportunity for biological evolution, which is propelled by this enemy of death and its scythe, that never sits at its feet, naturally or artificially selecting dead things as living beings. Indeed, a continuously regained space of light, as a magic candle on a birthday cake, which after being blown out comes back to illuminate again.

Or put in another way, instead of *resurrection*, maybe *return* or *rebirth*, always from death. That is, in the middle of nowhere a new existence appears, a new instant emerges. From event to event, from place to place, from heartbeat to heartbeat, from quantum level to quantum level, from galaxy to galaxy the Sun travels along a rope hanging between two cliffs, and after its deathly implosion in a black hole, not really knowing how, the *three* disappears and a *five* is reborn, the relaxed electron drinks a sip of light and gets excited. From the bottomless well, crossing an infinite unabridged chasm life returns, with no way behind the foot treads the ground again. Everything is always destroyed and created once again and new, and as now the First Law of Thermodynamics would say: that's how *everything is transformed*, without apparent holes, pauses or gaps. Thus, somehow, the First Law of Thermodynamics and death is convinced and defeated respectively; in this way continuity works, the arrow reaches and impacts the target and the hare catches up with the tortoise which had started the race first. There is not time or space for paradoxes.

Having said that, is this also the final message of the Gospels? Is Christ's victory over death what originates and leads to change in the universe? Weren't there any changes or transformations before him? Was his resurrection a qualitative jump, the birth of a new Adam and Eve? In short, is his resurrection or any other resurrection also an analogous symbolism to express the continuity mechanism within a discrete universe of separate pieces, of a broken universe? Is resurrection the origin of life, life that emerged from death, from non-life? But, was the resurrection of Lazarus the origin of his life? Does resurrection imply creation too? That is, does it imply not only the continuity of time but also the whole new moment of the present? But, is a new moment really possible? So, is life a daily origin, a daily resurrection, a daily creation? Or is it always the same, always repetition and past over past? What is one's life like? *One* over *one* until *three* becomes *five*? On the other hand, is reincarnation, already widely expounded in classical Greek culture and also in other cultures, another figurative narration of the same fact? Especially when it is understood as happening now, in this present world of quantum, discrete, punctual or digital nature? What if the world extends beyond the future, in the domains of final and eternal death? Nevertheless, doesn't a stone follow this mysterious mechanism too, from the bottom of the sea to the top of the mountain, step by step skipping over holes bigger than the stone itself? Then, is this the heartbeat of the entire universe, systole-diastole, death and resurrection? Is it just in the observable universe, the one which its quanta of light still reaches or could reach the observer? Will it be possible to resurrect and not go faster than light? Anyway, is continuity a fiction, like in a film made up of frames? Or are the frames the only fiction, the limitation of the biological eyes and the projectors, the essential limitation of the limited machines? Is life a complete event, without parts or sequels, the Nirvana that finishes with reincarnations definitively? Or is it an unfinished symphony cut up at its end instead? In this case, is the composer composing this symphony intelligently? In fact, a symphony in which the last inevitable silence will be part of the rhythm and melody too, finishing and completing it as one more note of the musical artwork?

Let's take up the abiogenesis studies and its limitations in this context of life and death again. So, it would now seem appropriate to describe a fictitious scenario situated in the past, a brief story slightly more dramatic than most scientific conceptual descriptions; that is to say, more dramatic than those insignificant emotional dramas where molecules react, bind, separate, repeal or attract each other, resist and release, and so on – material dramas where a pair of naïve teenagers seem to be acting. Thus, describing some moments where matter and light, the energy that is transformed through an informative or non-informative process formed a single superstructure never seen on planet Earth before, even though it may not have been its first cradle; a new structure playing around here, a new thing that would go down in history and therefore with the ability to make history. Hence, something with the ability, still supernatural and scary for many, to disappear or vanish completely, to fade away, to pop off, to finish like the flame when the wood is consumed. Or rather, one respecting the existence of the rope that remains without a knot, to allow the end like if it were the vehicle of a fatal accident, enabling to terminate, decease, die, go and pass away. So, using the common ancestor to stage this drama, that little old boy introduces the ability to escape and get rid of this deterministic universe of laws and submissions, dust, energy, mud, clay, chance, incipient and cruel nature. Was it a masterstroke? Check and checkmate, stalemate? Checkmating the universe, knocking it out? A goal scored inside the material net? Flip touchdown into nothingness? Was it just at this time? Was it the key or the crucial moment? Was this the origin of life? Is there life since the time something was formed that could or make it possible to die? The instant when the first fate or sense was forged? But, does this capability exist? Is there any sort of fate or sense? Is everyone sure? Does this confidence mean some knowledge, some wisdom? Or does everyone still expect or require some experience to confirm irrevocable facts, truths? Does truth need confirmation? What would be the foundations or axioms to confirm it? Or as chemists usually say, what flask can contain a universal solvent?

But, did the common ancestor really die? Is the journey that started finished? Isn't there life, branches, flowers and fruits from the same stem, from the same roots? Is its personal evolution still going on? Will its life exist forever in a relentless and unstoppable conquest of the entire universe, of the dead universe? Why didn't the common ancestor die immediately? Why continue if it had achieved an ending, the end itself in front of its young eyes? Was it waiting to get tired and suffer so that it could rest? Was a living being unaware of its own achievement, does it think it was everlasting while cloning itself, immortal? But, was it its particular achievement? Did mortality awareness emerge later? When exactly? Did the ancestor give a generous present when it found a way to share the game with death? Was it a poisoned present instead? An act of compassion with matter, dust and mud condemned to eternal slavery? A present for the golem? Was compassion the origin of multiplication? Was the original cell division an act of altruism, bequeathing and sharing the fate of death with love? Can the salvation that is delivered by Christ to the world through his death be understood in this sense? Indeed, the symbol that expresses or the history that narrates the escape from the dust that one was, is and inevitably would be? The fact of escaping from matter or energy whatever its form, from all the dust that constitutes the space-time of the universe? Is this eternal life, the absolute liberation of yesterday, today and tomorrow's omnipresent dust? Is eternal life being free from the only possible existence? That is, being free from a slave's existence tied to the gear wheels of the universe, to mechanical life? However, is this freedom an exclusive fact of the hereafter, always delayed? Or can it be realized by living beings here and now? If this is the case, what must they do with their mechanical bodies and existences? If this is not the case, why is there a delay and why is there a veil of slavery?

And now, returning to the stage of biological evolution towards complexity, is dying in a more complex manner the reason for evolution to go towards complexity? The reason for giving birth to more children and in the meantime launching competition against them, automatons with insuperable abilities? But, can one live or die more intensely, more complexly? Is the relationship between life and death an obvious correlation or only when it is understood and felt that life and death are happening simultaneously in oneself? Does natural selection select among the lively living beings so that they can die more profoundly tomorrow? Does mutation and drift lead to more dignified deaths, to more and more worthy living beings? Does death relish and enjoy its visitors? Is it a cruel foodie itself? Anyway, is life evolving to keep on the process of cessation of present living beings? To keep the dying and living process, day by day, eon after eon, moving and never stopping the wheel? Is it also the Samsara wheel, the wheel of life, the continuous flow of death and reincarnation? Is it also the ancestor wheel, the extended wheel of the I or the ego? Then, does evolution allow more different genotypes and phenotypes to die? And in this way, as if a door were opened in hell, the more the merrier, will more new individuals be able to leave this world of slavery? From nothingness to slavery and from slavery to freedom? Or simply, from eternal slavery to eternal freedom? In short, to allow anyone to be completely free from being anything more than a used machine or an ancient law? However, in this evolution process, what is what lives and what is what dies? What or who is released, what or who is taken by the hands of death and resurrection along the thread of continuity? Do they only take nothingness without hands? Besides, is death freedom or is it rather a sentence? Is it a sentence that condemns being free? Then, who or what is condemned? Perhaps, does the sentence mean stopping evolution, being the same thing forever, the same good or bad stone? Otherwise, dying or being forever anchored to lived time, to the dusty quantic field? Again, tied to the wheel, to the ancestor evolution, to the little ego that emerged and goes on?

When words are clean after getting dirty from saying them over and over again – even with the only hope of spreading dirty dust over sickly asepsis and programmed automaton minds that are fooled into slavery – life and death seem to have more than one option for understanding; moreover, depending on how they are dressed, they may be exactly the same. So, among other options not mentioned here, on the one hand, maybe life and death are two facts which are not related whatsoever and then mutually exclusive, especially if one understands resurrection as a synonym of life and consequently the radical negation of death; that is, one the immediate destroyer of the other, as the music or the noise that inevitably displaces silence. On the other hand, life and death can be understood as part of the same thing, practising or doing yoga in the lotus position; two dependent events that happen together, in unison, hand in hand, with one voice, like the two expressive eyes of a single face. Indeed, only a single word; or rather, a two-syllable word with a single meaning, a unique glance.

Thus, reviewing the facts before the complete separation or the total union of life and death, one can intersperse both in speech as in a mishmash, since duality and dialectics always seem necessary to communicate something about the issue in a more or less intelligible way, as the inventors of switches and binary code programmers know. Hence, immersed in this programmed appearance of the world, in the simplest reality shared by most observers, death and life, to die and to resurrect, are commonly regarded as an antagonist pair, a successive and exclusive or a simultaneous and interrelated pair. In both cases, a conceptual construction emerged in a dynamic reality, singular or plural, particular or general. So, perhaps already mixing both options subtly, the old master Laozi said, or at least, as it is quoted in some modern translations and interpretations of the *Tao Te Ching*: "life and death, only growth abstractions"; meaning by that, separate, simplified and isolated pieces that the passing of time peals and abandons. Only tools for conceptualizing change, movement, the true dynamic and creative reality, the Tao. At least, until the definitive and ultimate death, being the last step in a career that began the walk of life, does away with everything that started; definitely stopping change, growth, movement, the same dialectic between life and death, the relationship between the right eye and the left, the glance. In conclusion, one would say that before concluding a dialogue continues, being a cadence, a relationship that one of the parts – death – seems to end with the last word, a deaf-mute word, the eternal echo of its voice.

However, if one pursues the same matter, which is still thought to be unresolved, still thrown into confusion, can one base this relationship between life and death in the Uncertainty Principle of Werner, the one that states that *the one is more alive the other is more dead and vice versa?* That is, the more one talks the quieter is the other? Or are both speaking just as intensely at the same time? Perhaps, do they use different loudness, pitch and timbre? Different languages, different jargons? Do they speak without listening to each other at all? Are they perhaps maintaining a harmonious dialogue, a musical dialogue? Or rather, a dialogue of different interests, two monologues exchanging nothing, a fatuous dialogue? A dialogue between two talking parrots? Are both trying to ostracize one another by means of an accusing and reproachful silence? Otherwise, could it be an instructive lesson between a teacher and a student? But, who is the teacher, who is the student? Is there a lesson, something to be learnt or taught?

In any case, when eternal death comes to stay, why is this last cut different from others? What is the difference between definite death and all the previous deaths? That is, all the previous cuts made between each moment of a quantic life, of a life made of separate memories? Aren't there more resurrections at the last moment? Are they out of stock? Then, after the last heartbeat, does one die exhausted on the shore, a last diastole going on forever? Why isn't the magic candle that has been lit before so many times during this digital or discrete life re-ignited? Why doesn't light return from the confines of non-existence? So, at the last moment, when the definite and eternal death started, did Christ die as everybody dies, just as Lazarus ultimately died? Was Christ taking part before dying in a resurrection that was not his known resurrection, taking part in a natural or universal resurrection? Therefore, is his historical resurrection only the synthetized and narrated conceptual expression of his life, of life itself, of any life? The echo, the footprint, the fragrance alone that emanates from his life? Is it the Holy Spirit too, the memory of his health? The consequence of his life over history, just an autonomous voice that is still speaking now, just a wave that goes through space and time while transforming it? Just as any other voice, just as any other wave? Just as any action that influences its surroundings? And hence, without any possible alternative, when the very end comes, will any actor and its indefinite prolonged act be forever dead too? Or will there be any kind of reincarnation, any kind of re-information or reforming process after this definite end of everything known? Is there also a definite resurrection of dead beings? Will this be pure and eternal life?

On speaking about the relationships between life and death one should also mention, though only to be polite and respectful towards life sciences, the role played by death in the theoretical body of traditional and academic biology. Namely, death is not a word constantly repeated in biology books. Hence, it is presumed here that there is essentially a mutually exclusive standpoint between life and death; and when theorizing, death is what is excluded pitilessly, marginalizing it to an absolutely ephemeral event, as a hidden and almost embarrassing fact, as if it were apologizing for existing. Thus, its presence is avoided; and therefore, theory seems based on a continuous and endless resurrection that is not named whatsoever, rather it seems that it is silently taken for granted. All life is taken for granted and it is understood without a pause, the biological history takes place undoubtedly in continuity. A continuum that does not need a review or any inquiries, with no hole to fill or cover between its moments, neither between the inheritance of the genes nor between parents and children; it seems impossible to introduce a finger in any crack, wound or gap nor there is there a veil of nothingness making it possible to distinguish, to hug or to be a personal form. Consequently, when death is too near to be ignored, it always appears dressed in a furtive way, with very little intensity, like the light of a waning moon that shyly lights up a city at night but that has already been completely illuminated by life; although the light of this life is the artificial light of a borrowed fire.

So, only occasionally, between genes and biochemical reactions, the black shadow of death seems to appear behind a small and hidden slit, perhaps out from the place where it remains detained or imprisoned. Before theory forgets its existence, it appears from the back and crouching, when it talks about the relationship between living beings and the environment; when the ecosystem and food consumption, the food chain or net, the big fish eating the small or a lot of small things eating the big one is explained; when a cell engulfs another, when a bird dives into water and swallows a crab; in short, when one expounds the natural law of the jungle to a privileged part of the jungle living beings. Also, for instance, in the description of defensive mechanisms, as in automatic cell suicide or in the sacrifice of the combat elements in community groups while defending the territory or the mother queen; or simply, when an individual or a group is destroying the life of others, or one's very own and external life as cancer. In conclusion, in those cases where the body is preserved against attacks, illnesses or accidents; equally, when the same body forms part of these attacks, illnesses or accidents. Some examples of this could be a war between cannibal tribes, a bacterial or viral infection in the heart of a relatively lethal parasite or a scared rhinoceros escaped from a circus that tramples and crushes a deaf and half-blind cat that was crossing a road while the traffic light was red.

On the other hand, with somewhat more prominence, death appears when talking about evolution, because the latter relies partly on it and then makes it possible to shape its face superficially, as well as just indirectly, according to causal effects alone, similar to a black hole in outer space. Thus, the entire evolution is based on death since it can only be understood through the change that relies on the end of something, the end of any state whatsoever, of any shape, behaviour or previous biological piece or system. Specifically, this is more evident in the theory of evolution by natural selection, which suggests that death is a sort of indirect designer without positive action, guiding the forms by only prohibiting the transit in its own possession, like a big rock in the middle of the river that alters the course of the stream or like a filter or a meat grinder. As such, death is the limit of the design produced over time, a design that is unable to adapt to the present flow; for example, the end of an allele transmission and thus the end of a particular physical trait that will not be repeated: extinction itself. Therefore, if it were the end of a design made by a designer, at the same time this death would be the dramatic failure of the designer himself, who was unable to predict the future, the new environment where his invention, artwork or experiment has been put to the test. Or obviously, the precise success of this designer, who wanted exactly the end that occurs and hence who knew about the emerging environment. Anyway, in such a context, death almost shows the skin of the nose, but still remains an unexplored habitat for biology, since it remains still in the unknown future, beyond the pages of the academic world and experimental reality. But, why is thanatology the poor sister of biology? Is it a sister repudiated by her family? Does thanatology only care about legal relevance, forensic investigations, clinical deaths specifications, incinerations, coffins or inheritance management? Is it only the winner that has the upper hand in biology? Is the living being the winner and the dead the loser, the one defeated? Is the loser perhaps also the winner? Who can understand and accept this last statement? Was the whole aspiration of Christ, just like everything else, finally defeated too? Are all the dead livings in history utterly defeated? Is life, except the present, part of the everlasting defeat? Is this perhaps the absolute victory? Is that a sort of mysterious victory, something that is difficult to understand and comprehend for everlasting winners?

Once attached to this influential tradition used to compose the biology field conceptually, also used to compose much of the cultural modern world and its media – that is, being conditioned by the ubiquitous dialectic between victory and defeat, success and failure, the will to power, the fratricidal and perpetual competition between living beings, and so on – inevitably escapes from a large number of texts and television documentaries, academic talks or books on the so-called nature, a world where, for instance, the impala and the lion live together regardless of that. A real world, an actual moment where they both look at each other's eyes as if they were without a care in the world; with no hunger or no wish to run, with no threats or fear, totally absorbed in their own affairs, infinitely enjoying their own bodies and the Earth. Being a lion and begetting lions and being an impala and begetting impalas. Perhaps, feeling inside a fleeting spark of compassion for one another, understanding what the other has to go through, realizing each other's misery; because, one must chase and the other must escape, both yielding to the demands of life, the need to work and earn a living. Also, maybe being curious about the other's recreation, imagining what the neighbouring leisure is like; the required neighbour around whom an entire life is developed throughout time, mutually influencing each other while a common future is built, perhaps of less needy nature.

So, at least in the savannah at noon, when the Sun is scorching the leaves of umbrella trees and invites everyone to sit for a while in the shade of its minimum territory, the day rules over them; this solar world or any other equally domineering, like a harsh winter that shows everyone the way to the shelter or a hypnotic spring that has an enchanting effect with its fragrances, allows lions and impalas to drink the joy that flows from its plenitude, from its health. A world letting lions have time to digest the food slowly and letting impalas exist outside the stomachs of lions while grazing the still grass that does not escape at all – living in such a different way from impalas that perhaps they feel that the happy-go-lucky and non-muscular beings do not deserve the privilege of resting. In fact, a very real world, always present but in continuous motion, which in a way expresses the overabundance of life in time and space, with more than enough krill to sustain the whales and zillions of ants and termites to feed the birds and their offspring while going on with their daily own business too. The overabundance of light that allows inefficient energetic processes as the photosynthesis to develop plants and other organisms and then to sustain a huge load of parasites on their backs, including krill, whales, ants, termites, lions and impalas. A world that allows the glorious rest of some cats on the roofs, peaceful sleep and not having to worry about problems. In turn, a world where it is possible to move towards an indefinite complexity, or at least, which would appear to have been possible so far.

And nothing seems more possible today than sleeping and thinking, although not so lucky as the cat on the roof. So, sleeping and thinking, overabundance is tangled with life and death in this light dream, thus one may wonder, is overabundance life itself, its unique quality? What is the origin of overabundance? Is this also the origin of life? What does this overabundance imply? What is its source? Does it imply that the oceans of life are much deeper than the shallow puddles of death? Is death absolute scarcity, *overscarcity*? The open space behind the door of hunger? Why is there always hunger among natural overabundance? Why does it hide behind the excuses and theories of competition? Is competition itself what causes hunger? Is there any counterweight to overabundance, preserving some sort of balance between life and death? Who or what is always hungry and dying? When or where is the surplus totally consumed? When or where is one unable to receive it? Who or what dies in an overabundant world, made for life and for living, full of life, full of protein? Where and when isn't there more life?

That being said, one wonders for a moment again, as if one were reading a book: will I die? Will overabundance let me die too? Will it let me go from of its hand of plenty? But, do I die now, am I dying right now? Am I asking this without any prejudice, can I ask myself as if I knew the answer? Can I ask this and believe what I reply? Am I answering this, can I answer something to myself? If so, is the speaker different from the listener? Are they one? Am I both? Anyway, is the I the thing that dies? Am I the I that dies and will die? Who or what is I? Is it just something that can die, capable of dying? Is the self an invaluable object with the singular ability to disappear completely, making everything special, too special and terrible? Is it the only known thing in the universe with this capability, the goal and the conquered treasure of life? However, do I, or my I, myself, me, have this ability too? Or is it special, unique? Is this an innate ability or has somebody taught me or has yet to show me something about it? And furthermore, does the cat that is sleeping soundly on rooftops have an I able to die too? Can the little and humble I of the cell die too? But, is a little and humble thing the I of the cell or could it be as big as the I of an arrogant and vain whale? Is it possible to measure the I, to ponder over the self? Then, which would be the unit measure in the international system? Is it possible to compare the I with something else? With a universe, with a stone? What about the enigmatically silent I of the stone, can it also die? And the third person, and the single and the plural you, and we and they? Are the personal pronouns these naked beings that transcend matter, energy and all mechanical laws, the actual rebel laws? Or on the contrary, are they the most profound manifestation of these, their iron and nickel nucleus, their Praetorian Guard? Is there anything that belongs to someone? Can life be lost? Is life a possession? Who is the owner of what is carried away by death with or without permission? But, is the whole life, the whole being, the whole living, the whole living being carried away? Simply, is it the owner who is carried away? Therefore, what was the body that remains, a wallet full of money that is totally empty now? Is the end a robbery or is a bit of justice for a thief, for a defaulter? Can one experience death in life and see what happens? What happens when an everyday event finishes, what remains, what begins? Does it come to pass what the

Gospels proclaim, what the *Quran* in Arabic says? Just as any other book that expresses some indefinite continuation of life after death?

Finally, flying like a vulture, going round in circles over a dying body, one can describe some arbitrary end situations based on current and simple experiences; or perhaps, describe a little fantastic story that represents eternal death superficially. Anyway, a story that tries to end without a second part, let alone a trilogy. Then, one can say that the end, death, is revealed fully when there is no support, like a foot while expecting to tread on a solid surface finds itself stepping into a hole. As if Neil Armstrong, when leaving the spaceship ready to walk on the Moon, after his entrusted jump at the hands of lunar gravity, had not stood firmly anywhere and had gone on weightless, floating in the same place in space or drifting, wearing the same spacesuit forever, always following the fashion of the sixties in the twentieth century. Or also, in another more extravagant and wilder scenario, another astronaut, leaving another spaceship much farther from his native planet or its satellite, had just set foot on the field of vision of the eye pupil of a black hole, which would have swallowed him immediately with no time to talk nor to hear his first words, stripped naked much faster than it took him to get dressed in the first place.

So, when definitive death or the definitive end comes, there is no adequate support where the vital processes can be carried out. No moon to tread on, no body to make it move; moreover, without a useful brain to control, choose, think, imagine, record or remember; neither to sleep deeply nor to dream or trust tomorrow. That is, without any time to spend or to count and without space where to shift; without a mind or a world where being I, you, he, she, we, you or they; without pronouns with whom to identify and share non-existence. With no place to be something, without ink to write and with no life to live. No more or no less, everything lacking, no existence to contemplate or death to face, also deprived of the ability to die, the treasure of life already spent. However, just a minute, is there nothing left where was before? Does anything survive from this precise and devastating incineration? Any support, a boost for a dependent ghost that cannot stand for itself, a fire-proved bed sheet? Is there anything left to resurrect after this fire that breaks up the configuration of life until losing all that is familiar? Is there an amorphous void left? Will there be at least the open arms of death? Will there only be eternal death in the place of eternal life? But, who or what is death? Is it possible to identify oneself with death to live its life forever? Is it possible to travel to its den, to enter the heart from which nothingness emerges and then to take its place and sit on its throne? Have any living beings taken the place of death after their own death? If so, is the void already filled, with no hole where to fall any longer, as if it were filled with soil to walk over it firmly? If not, what or who could carry out this mission, who or what can fill a gap of that size? Who or what is more than all the dresses that death can easily strip away? What is the dress that is not a dress? What is the dress that does not dress? What is absolute nudity? Who is so whole and firm that the death scythe is not able to hook and pull away, as if it were a slippery body without handles where hooks and blades slide and slip? Is oneself the only in charge, the only useful for that kind of missions?

Otherwise, although it is almost the same story, one can imagine another dramatic setting where the end is staged simply by cutting or untying strings, like a cage or a prison that opens its doors and a bird or a prisoner flies away from the inside. As it has been aforementioned, the end of slavery and the beginning of absolute freedom. That is to say, the Nirvana, the Kingdom of Heaven, the Janna, the Promised Land, the promised planet, the street viewed from the prison, the Father's House, the end of a mortgage, patients' health, good news, and so on. Well, it seems that nothing can bind the dead being. It does not have a body to worry about, no name or address where to send invoices or bills, it does not have to drink or eat to maintain an organic structure that fights against the current like a salmon. It does not have thoughts or any need to sleep, nor carry out any other vital function. No needs at all. It cannot help or hurt anyone. It does not talk, dream, play, pray, go wrong or sin. It does not have to take part in any competition. It does not suffer any pain or feel any pleasure, desire or fear; all wounds have healed and the scar has vanished like smoke in the air. All things have been forgotten. It does not have to work nor have to rest. It neither lives nor dies; it exists no longer, it is really free from everything it was, it is absolutely nothing.

But again, who or what is released, who or what is free? Who or what abandons the world of need and fear? Who or what is now nothingness? Is this transformation possible according to the old First Law of Thermodynamics? Is it possible to speak about nothingness or only silence can say something about it? Beyond *who* or *what*, does the concept of pure freedom have any meaning? Does something like this exist? Does it denote a reality, a fact? Can one know something about its existence, see an example or taste it with the tip of the tongue? Has anyone experienced or heard anything credible and honest about it? Meaning by that, a living being without impositions, without pressures, without basic needs or acquired vices to satisfy, with no debts, no responsibilities, no charges, without hunger or sleep, without any dependence of any kind? A living being similar to those unborn? Someone who inhales the air of freedom and exhales words that tell the others what death is like? A former prisoner in the street? An emperor or empress doing what they please? A son of oneself? An ideal child playing absorbedly in an ideal childhood? A mature adult living life to the full? The parents watching over their children lovingly, living for them and with them? But, aren't kings afraid of losing their kingdoms, aren't parents afraid of forgetting their children? Is fear compatible with freedom?

Otherwise, is it possible to find something or somebody free in a radically opposed way to that of absence or indifference from dependence and suffering? Also radically opposed to mystical moments, moments of pleasure or happiness? Is it possible to free oneself through the absolute acceptance of all kinds of slavery, debts and charges? A sort of freedom that continues to deepen in the mud until finding death hidden within life itself, as if it were a stowaway or a saboteur, glued and tangled with life like a sticky parasite, like a banyan or a tick? What would be this kind of freedom? Is it the freedom of someone free or is it the freedom of surrendered slaves? Is this realistic freedom, the freedom which bears the real world in an objective mind? Then, is this unconditional freedom? The freedom to shun choice between alternatives, since there is no real alternative to some kind of slavery? Is this the truth that makes one free? By means of quenching the desire to escape, to escape from suffering too? Ending with the desire to die, with the desire to be free, happy or lucky? The absolute resignation to all self-will and starry-eyed dreams? So, willingly integrate oneself into any form of slavery or obedience? But, isn't it the substitution of one form of slavery for another, being a slave of the outer universe instead of a slave of the inner I? Who is a crueller tyrant, the selfish I or the universe that is a stranger to oneself? Would one innocently expect anything positive from any of these two cranks? In any case, letting oneself be chewed and digested by the hungry lion that has to feed their cubs? Getting involved with overabundance of life, food, strength, number, and so on? Being the highest expression of overabundance, light over darkness, the source of overabundance itself? Whatever the path one takes to freedom or the absence of any paths, can anyone understand the ultimate and eternal death through this example? That is, definitive death as a synonym of absolute freedom, freedom of continuous resurrection? Is it more difficult to understand for some than for others? Why? Nevertheless, does everybody die but is it not the same for everybody? Or is it?

5

Inside the old dead bark of the tree,
the sap runs through it quietly like a happy child.

Ψ

PEACE,
probability and the multiverse

When estimating the accurate probability of the original manifestation of biological life and what remained subsequently from then until now – with all its peculiar details that are known and are becoming constantly known – from a lax mathematical background and without requiring for it a sort of *attractors* or something similar to gravitational or information pools that lead preferably to something definite – such as a black hole, a gene or a single word with or without an apparent meaning – also renouncing to define where the probability of the impossible begins or ends, and leaving aside the immense and crucial error that can be made only by separating the lips of the mouth an electron or a photon, in this context, the estimation process easily

114

reaches disconcerting conclusions long before one watches over the beginning of the end of these calculations. That is, one starts to write very long numbers, extremely low or tremendously high; anyway, numbers that do not fit completely into the meaning of the word *oddness*. Besides, if one tries to write them, there are neither enough oceans of ink nor sufficient paper in the world to express them without artifice, to show them fully extended, unfolded, without the use of misleading exponentials, not even using the art of nano-metric miniatures to represent them.

So, one often tends to ascribe or assign life success to a negligible probability of occurrence. For example, although it is a rather long example, considering a number below the number which represents the probability, or above the inverse thereof, that the protagonist of the *infinite monkeys theorem* – that monkey that with an infinite amount of time would almost surely write a Shakespeare work tapping on a keyboard at random – should write with a reed pen or a duck feather both parts of *Don Quixote*, soaking the tip of the pen in ink periodically, writing in a stylish hand from the first to the last letter, in a single row and with the spacing included; of course, once overcome before, with two glasses of wine and a ribbon that ties the pen at its hand, the natural rhythms that makes him or her a monkey, that makes him or her an it, and so on. Also, since one wants to present an unusually low or high number, doing it with the same type of handwriting and with the same signature stamp which appears in the original autographic manuscript of Miguel de Cervantes; with the same food, drink, sweat, grease and oily powder stains which almost surely appeared before being edited, with the same folds, creases and flaws in the papers that the copyist received for its printing. Ultimately, the number that represents the exact probability that this monkey really exists and it does exactly what has been said, on the first attempt, on the first roll of the dice and right now; obviously, a number that is lower than the number that represents only the probability of a monkey simulation doing this theoretically one day. Thus, ignoring the probability jargon and mathematic formality whenever one wants, is the probability aforementioned an infinitely low number, is it zero? Or is it a zero followed by a radix point and an infinite amount of zeros behind, but finishing with a one, sad and alone, emerging from error when rounding up? Is it an impossibility or stranger things have been seen? Can infinity grow infinitely, be infinitely higher or lower, infinitely great? But, is infinity real, is it a thing? Is it a fact or a speculation? Is it a fixed and already established thing, something dead, as dead as the word *infinity*? Are there more impossible things than others? Is there a limit to what was or is heading innocently towards infinity, something like the limit of the speed of light in a vacuum or a certain entropy value at zero Kelvin degrees? A sort of stone wall where all hope or illusion always crashes? Is any real limit also some kind of infinity manifestation, such as infinite barriers, insurmountable heights, bottomless depths, unbeatable enemies, unfathomable realities, incredible lies? Is the limit the infinity upside down? Is it the face of infinity instead?

When numbers go beyond everyday experience – tired and asleep long before counting sheep from million to million to googolplex – but at the same time it is observed with astonishment the successful and orderly outcomes of a process that is meant to be forged with the hands of denial and chaos, as if a dangerous and ugly weed was growing in a beautiful garden planted with lovely flowers, in these reluctant cases, the word *miracle* is often insinuated and also avoided by means of recalculation. In fact, is it more probable to deny a singular life than to accept the most common of miracles? Is this the common miracle? Is a miracle the rebellion against universal laws, would this be the basic miracle? Is death or the incredible ability to die the terrible and glorious miracle, the usual and scary miracle? Or rather, is a miracle simply the favourable success of the lowest of probabilities? The probability that represents unimaginable numbers, bulky giants or tiny dwarfs unable even to tread flat ground? What is the probability of being and what is of not being? Is it a fifty-fifty as Hamlet seems to estimate while dialoguing with a skull? Fifty-one against forty-nine, simply because Hamlet speaks and the skull keeps silent as if it were a stone? One single favourable case facing infinite possibilities? However, was the existence elected in a referendum, was a democratic election based on percentages? Is anything a miracle for those who are not familiar with probability laws and distrust the numbers they cannot count on their fingers, neurons or using all the physical particles or fields of their brains?

On dealing with these miraculous cases, it seems appropriate to reconsider and delve into the computing ability, also into the expression, understanding and comprehension abilities. The role of limitation in all this business; the relationship of knowledge with the longitude of the mother tongue, the size of the ears of the listeners, the degree of blindness of the observers' eyes, the capacity of memory of living beings and their current genes, and so on. By way of an example, the huge limitations that can always be surpassed when one has all the time in the world, as in the case of the monkey of the mentioned theorem, but that inevitably arises when trying to draw the attention of a snake or an ant, dead or alive, even if one only wants to convey trivial information about oneself – one's favourite colour, the opinion about the so-called politicians, etcetera – even if one only wanted to inform them about themselves, explaining how the forked tongue smells or the antibacterial properties of formic acid – its chemical mechanism of action, and so on. Also, as another example and using certain popular culture, the limitations of lichens, cacti or chickens to enjoy or criticize gracefully the cubist art of the *Woman with Mandolin*, the philosophy of the *Critique of Pure Reason* or the aria of the *Queen of the Night* in *The Magic Flute*; otherwise, the limitations that Pablo Picasso, Immanuel Kant or Amadeus Mozart would have had to develop and share their work while inhabiting the body of a shark or a sewer rat, even though they would have tried to share their works and ideas with the most savvy sharks and rats of the respective species. Or above all, the limitations that the singular automaton will have in order to explain to its hominid inventors what the technological singularity and the new ultimate meaning of life are.

In this almost perfectly understood context of temporariness or eternal limits and restrictions, which include the memory of life being now partially preserved and expressed through genes, it seems that it becomes necessary to focus on the thought that emerges from this memory as if it were the shape of a ghost in a bed-sheet – an extended sheet that is vastly shared, the ground that almost everybody steps on or the Sun that rises and sets for everyone, at least for those who are out of bed playing with the sheet under the sky. That is, focusing one's attention on several activities of the neural configurations capable for millennia to ponder, think, represent and convey with some success something wrapped as a present or as energy in the quantum world; for instance, words or concepts such as *paradise* and *evolution, time* or *madness, hell* or *happiness.* And then, at the same time and in the same act of communication, also capable of sowing and nourishing these seeds of thought as if they were rational bread rations, until they flourish among the more or less clever brains of humankind; in this way, finally organizing their neural configurations to allow the repetition of this process over and over again, as humanly possible, in all potential directions, as if it were an epidemic. So, despite its being possible to understand or sense rather quickly, almost instinctively and only by means of a brief tailor-made explanation, for example, what *karma* or *sin* are, could the same thinking and communicating process get a clear idea of the name of the variables which it uses to calculate some probabilities? Perhaps, the number of stars in the universe or the distance from the two most separated? Do these huge numbers have particular and familiar references or are they part of a shapeless abstraction that encloses all inconceivable stuff in the same bag? Would it be a miracle to achieve a detailed mental image of this immense stellar scenery or is it commonly achieved without a need for a miracle? Anyway, is it possible to compute the total number of neutrinos in the universe one by one without getting bored and make fun of a neutrino per each Planck time unit that elapses relentlessly? How many automatons would be needed for that, how many drunken monkeys are needed if one does not want to sleep awaiting the result of the sum? Incidentally, would it be necessary, among other things, to compute all elementary particles and antiparticles of the early universe, and all their possible configuration settings, if abiogenesis wants to develop and run a numerical model of the universe that simulates the key moment of stochastic life emergency in a precise way? Could this model use wider concepts to save such tedious work? Or use some kind of tool that already considers and truly represents part of the past history as if it were a base camp to climb to the mountain peak? Could it be the brain and its mature abilities this base camp? Or is the brain the mountain that other brains try to climb by using theories, thought and its assimilated words as a base camp?

Once again, some of the problems can be reduced to the same complexity of the problem that is brought up. Hence, it does not seem possible to compute, understand or conceive infinity with a finite and limited brain, nor with any other mechanism or alternative hardware or interface also finite and limited; no matter if knowledge or understanding readily accepts any words in its approved dictionary. Obviously, it neither seems possible to reach the largest infinity nor the infinite infinitude of infinity, with all current brains and computers at full capacity and networking as if they were a single brain or computer. Not even by making this brain a great mind, an interactive psyche following far less restrictive laws than the known physical or psychological laws, even if it is able to dream about such strange or mysterious things such as an eternal infinity of blood, bones and flesh. Nevertheless, can the heart in love help to deal with this endeavour or does one always need a finite revelation emitted from infinity? But, what is love? Is an infinite revelation possible? Could it be understood or comprehended or would it be like a background noise, like the sound of silence? Is infinity only a concept that means too great to count, too small to indicate, too complicated to understand? Is the expanding universe itself an actual limit that makes actual infinitude impossible, the potential limit of all things? Is the universe finite or infinite? Can something be infinite with a beginning? And something with an end? Is there possible infinity within a closed and occupied volume, maybe with an inside exit towards the outside? When those huge numbers that can be expressed by means of exponentials are they part of fantasy and science fiction? In any case, could any finite number, a number with a name of a number, be without infinity from which to take a discrete quantity and to keep growing? Are discrete parts possible without the unlimited, without infinity? Would there be a limit, a face or a form, without an infinite sheet that folds and takes up a shape? Must there be an excess in order to give away? If when taking or borrowing from infinity there is always more and more and it never seems to run out, is this a miracle? Is overabundance a miracle? Is the need to eat infinite as is the value of the king in chess?

On the other hand, without going so far and going somewhat back, is it possible to understand the unit? Is it possible to understand number one with a brain connected to the outside through multiple senses? Then, can one understand any computation of units? What about the zero without understanding absence, without reference to something that is not, without the previous existence of the one or any of its combination possibilities? Is zero a number or is it the space where the counter counts? If so, is time the counting process and zero the end of time? Is it also the beginning of it, are both the same? But, is zero a thing, something shaped, a pattern? Or is zero the only real and accessible infinity? Is it possible to be a zero? Is existence compatible with this number infinitely empty? What or who is a zero, who or what has no limits? Is the number of ones or sand grains of time that dead beings will remain dead a number above the ability to count indefinitely? Is final and definite death also eternal death? A borderless sand desert that transcends all the limits? Or is it more like a single grain of sand at absolute zero, arranged like a glass and motionless forever, always faithful to its narcissistic perfection? Otherwise, will the dead rest for only a certain amount of extra time, just for a little longer, until the hypothetical end of the universe and its own time? Will the universe and its time be born newly again overnight? Will this new universe be the definitive resurrection of the dead, the expected *Parousia*? Or won't it have room for them or their lives, perhaps neither for time nor space? Will the universe die but won't it have time to rest in peace? Will it resurrect instantly, like going to work without sleeping after an insane party at night? Will it rest at least a day, having a few moments of peace, relaxation and absolute oblivion to begin fresh like a tender rose? Will this be a day to remember, a terrible remembrance day? In any case, will it be reborn as if it were a totally new thing, something completely different?

Without shying away from numbers and especially from colossal numbers, as it is known, life sought through a scientific prism depends on these unthinkable numbers. And so, these are essential numbers too, truer than any other number, more real than any thought. Hence, the more one studies and researches while tying up loose ends, more and more is revealed, more knots and neurons are connected, making up a synaptic net that increasingly dresses itself with more knowledge. Therefore, knowledge – also knowledge about life – is always finite or limited, since always a longer, deeper and wider boundary can be drawn or built up, step by step, as if it were a safe wall where kids play while insecure and innocent parents believe their little ones are safe. Likewise, the wheel of causality is also redefined bigger and in a more intricately intertwined way; however, a wheel that always remains closed too. For instance, to establish a starting or cutting point, more or less shared and very simple, it can be said that *known life depends on the presence of liquid water*; therefore, the presence of liquid water depends on other variables too; and the more one knows about the mechanisms of formation of water and the necessary conditions to maintain it on planet Earth in this particular state of aggregation, more and more facts are pointed out as crucial, more specific events of all kinds which have been expressed homogeneously at times as digital or punctual events but somewhat related. That is to say, they have been put as a complex network of discrete events forming together a natural history, a narrated history which will never be the actual immense history, but only a sort of inflated word that shows a diffuse and partial footprint, knowledge itself.

Thus, skipping many more steps than steps walked, as the common study of abiogenesis does, some of the intertwined sequence of hypothetical events that caused the presence of the atoms that form the water molecules could be listed. After that, if this was the case, the main chemical reactions that formed water molecules, from these elements or from molecular reagents that contained them before, could be explained theoretically too. Then, there was the process of accumulating relatively so much water between the atmosphere, the continental crust and the ocean basins, and the succession of events that allowed the planet to achieve and maintain for millions of years a specific range of pressure and temperature in which sufficient part of the water is mostly liquid. But before going any further, is it possible to define the fully specified range of the inhabitable? The complete list of contingencies that allowed and allow life, listing all the necessary conditions one by one? How many things will have to be taken into account when designing an idyllic paradise for the automaton, a garden where it can go without a loincloth and without any worries? How many absolutely necessary things will go unnoticed during this ecoconstruction, allied or enemy of the objective, without knowing anything about them? Is it possible to reduce an inhabitable place to three or four variables with known names to most? Water, earth, air and fire? Is it possible to reduce it to a single variable? Affordable energy, exergy, work, clothes, good luck, television, power, obedience, phone, Internet connection, money or food? Will it also be necessary to programme building something similar to a saviour, a chosen one, a prophet or a messiah to feed the automatons with new words? Anybody who can give them dressed sounds, something or somebody able to shape and adapt the air for their powerful ears and brains? What are the conditions under which all life on the planet, included the automaton life, would inevitably stop finding support? A giant and fast meteor, a solid mixture of diamond and gold facing the Earth, would that be enough? A treacherous and unexpected tantrum of the Sun, tired of illuminating darkness and the thankless void of nocturnal hearts and dim eyes? Is it possible to build an automaton capable of resisting any universal phenomena? If so, as has been said before in other words, will this automaton be the evolved body of the current weak biological body? Just other different hardware? Another flesh for the spirit, language, mind, information, etcetera?

Nonetheless, after a successful attempt to consummate *Revelation*, once the Earth deserted or shattered with words or meteors, what is the overall state of the universe that prevents life from emergency again, any possible regeneration or resurrection? Would it be a second emergency of life after a devastating cataclysm a much easier resurrection to understand than others? Under what conditions, under what universal variables or circumstances, is the local emergency of life or the panspermia of the nomads on the outskirts impossible? When would infertile meteorites and comets land constantly on infertile eggs and soils? Incidentally, is the Earth an impregnated egg by means of an extra-terrestrial sperm? Do terrestrials have to start packing in order to continue the panspermia and avoid possible tragedies, only if it is to increase the chance of survival? Isn't this the basis of a biological dogma or imperative? That is, to procreate as many descendants as possible, moving as a mob over time until the machinery derails? Is this an altruistic dogma? When following the biological dogmas literally, what is this behaviour like? In any case, is it possible to change the conditions of an entire universe to make it inhabitable? Is it just something dead that could do this, just something that does not need anything beforehand? Then, without the necessary tool to do so, are there safer places than the Earth, more gentle and less polluted planets with more resources and possibilities? Is it reasonable to think that all the living beings on Earth could better suited on another planet, in an alien place where they would breathe more easily once grafted or implanted? Is it possible to manufacture a mechanical habitat just as a humanlike automaton is manufactured? An alien spa with the appropriate gravity to walk as if one were dancing and where all creatures ate as if they were in an intelligent restaurant? Is it possible to produce beautiful nature, attentive and without hostility, an incomplete ecosystem waiting to complete itself by planting a seed that will be cared for as a pampered child? Is it possible to build another mother Earth? If it is possible, must it also be programmed to monitor the changes that will occur after the inauguration? How will the automaton cybernetic mother respond to adolescence, maturity and old age of her grown up children? Will she only give birth or will she nourish them with her breasts and caresses too? Will she let them abuse and corrupt her for love? Which is the homeostatic programming law that will be applied? Will it be the unlimited or infinite sacrifice? Then, will she collaborate with the exploitation of her own resources? How many generations will be taken into account when programming and building this mechanical mother Super Earth? Will this automaton be able to sustain infinite generations that cry and suck more and more complexly?

In order to continue the journey along the inhabitable path that is shown initially by mothers – especially if human beings want to find new worlds to fertilize or new worlds where to take a stroll in the street with the same clothes – it is necessary to bear in mind as way of an example the privileged position of the Earth in the Solar System. Because, a little closer to the Sun as Mercury, then there would be too much light, too much heat or too much skin; a little further as Uranus, then the coat would crush the bones as if they were made of butter; in the same way, a little closer to the massive and gaseous Jupiter or a little further away from a smaller and more fragile Moon, and all on Earth would change or begin to change radically. Besides, leaving aside time dimension for the time being, the relative position of the Solar System in the Orion Arm can also be pointed out and subsequently its position in the Milky Way, the proximity to other bigger stars, the distance to supernovae or black holes, the trajectories of other large and massive bodies as meteors or the presence of the ultraviolet, X-ray or gamma ray radiation that go through the galaxy at the pace of a snail, and so on and so forth.

So, since the wheel of causality seems to turn indefinitely at one level or another, maybe it is better to stop it categorically and question the whole thing. Thus, is everything or almost everything that happened – and what did not happen – in the history of the universe absolutely necessary for the formation of cellular life, for the emergence of the common ancestor and its development? If so, does it mean that there is only one life in the universe? Or does the absolute contingency or dependence also allow vital multiplicity? Is there room for manoeuvre? Does life, without too much difficulty, grow like mushrooms in the field in innumerable fields all over the universe? In this case, is there any connection between these different living histories spread over the universe? Would they be truly like mushrooms in a wet and dark field, scattered and visible fruits of the hyphae structure of a hidden mycelium, of a single fungus? Or would they be absolutely independent lives, indifferent to other vital manifestations, too remote from each other? Would they be lives without any common language in which to communicate, not even English, Spanish, Hindi, mathematics, climatology, biology or Mandarin Chinese? With no information to share and recognize? Maybe, the sweet music of some would be the loud noise of others? But, is all life unique, unrepeatable and contingent? Or is it almost as common as gravity, as the existence of stars with planets orbiting around them? Does this last question become simplified or complicated when one wonders about one's own life? Meaning by that, when the I is the only subject of these musings? Thus, is the I, any I, unique and unrepeatable and all that happened is contingent or indispensable for its existence? If so, does it prevent the existence of other egos in some way? At least, does it make it very difficult? What is the difference between egos? Is there any difference among the various selves? But, what can the I experience different from itself to be able to conclude, knowingly, beyond it? Who has seen anything with the eyes of another? Who has felt something without being in contact or related to it? Where does the skin end and where does the air begin? Is this boundary made of matter? Is there any frontier or boundary at all? If not, what is an I or an ego without limits, boundaries or frontiers? Is it something? Is freedom the absence of frontiers and slavery the opposite?

When dealing with the contingencies or imperative needs of biological evolution, from the common ancestor to complex multicellular organisms, perhaps one notices new details of the relationship of life with probability; for instance, what is the probability of the formation of a group of apes removing each other's parasites under a tree – they, the tree and the parasites being descendants of a single original cell? Is this living scenario even more incredible than the first ancestral cell formation? Or once the ancestor was formed did everything go downhill without brakes, as a snowball? Is the probability analysis of the biological evolution the study of the architecture of a pyramid? That is, is it the stone lifted higher up when and where there is less area to build? Does one thing make up for the other? Or is it quite the opposite, an inverse pyramid? Or is it rather the architecture of a bridge perhaps, flat between two banks? What is the conditional probability assigned to a big piece of natural history? For example, from the first hydrogen atom to the conscious structure capable of observing and denying observation, to open and close the eyes to the whole universe as if they were a thin curtain of skin? Is there any lubricant or catalyst to favour and raise some insignificant and exasperating event probabilities? Maybe, is the probability of being conscious always one? Anyway, was certain ancestral information hidden behind a rock while having a Royal flush, infinitely eager to bet it all by means of storing itself in genes to make cells? Was the road paved and did the foot suddenly go into a tailor-made shoe, like the foot of Cinderella and the glass slipper? Did a spark ignite the coldest polar ice and then the entire planet began to burn spontaneously? How many past events were decisive and how many can be ignored in this burst history? How many things does life still depend on, how many things will future living beings depend on? And now, is there much more room to avoid getting away from the inhabitable interval? Is it possible to modify this range, making it so flexible to achieve the immortality of the weakest? Does current investigation and technological work mostly intend to achieve the immortality of the fittest?

Because of the tribulations that have arisen from the evaluation of scandalously low probabilities that reflect the delicate fragility and rarity that is assigned to some phenomena, imaginary or real – like a vulnerable, weaned and unaccompanied baby cell which triumphs in the middle of the universe of charges, masses and physical forces – some solitary thinkers who perhaps want something more than the solitude without boundaries, have expounded new hypotheses or reinterpreted some earlier hypotheses with new words, that fit with this narration now. One excellent example is the *multiverse* hypothesis, based on a useful concept intent on solving the confusion that arises from the extremely small or extremely large numbers, which cannot even be pronounced without leaving one without saliva and vocal cords. So, as the probability of occurrence and maintenance of life seems to be so ridiculous, even without considering the whole history but only considering the stage before the formation of a common ancestor and a bit of time and consciousness afterwards, then, maybe, it has been reasoned that this universe full of life, the historical observable and unobservable universe, is a sort of universe that won the lottery, a prized chance that was drawn among an astronomical number of tickets; besides, a raffle in which all the tickets were sold. Therefore, under the umbrella of this understanding, living beings are the manifestation of a single favourable case, results or fruits of a very low or minimum probability of success, for the sake of something similar to luck or fortune, or similar to a sentence imposed on someone innocent in a game of chance, like someone forced to play in a Russian roulette. So, life would still be enjoying a very big surprise, playing an astonishing hand in a game which seems that the rules and the incalculable number of cards allow countless combinations without a reward or a punishment, which would be the rest of the countless universes, the enormous number of other possible cases, other tickets or other hands, dealt to finish the card deck; now universes left to their fate like loveless pilgrims, each of them going their own way and looking at the ground, perhaps looking for coins or worms.

Thus, the existence of this universe with life is justified by the simultaneous existence of the pile of alternative possibilities; it is based on total diversity to rationalize this unique privilege since it takes all sorts to make a world. So, it seems that a universe of three Kelvin degrees at average temperature, almost transparent and tasteless, would be only possible if there were also the full spectrum of other universes, showing all possible average temperatures, opacities and flavours; other universes or other worlds, alternative or parallel universes. Perhaps, even with their own laws or with no laws at all, subjected to anarchy, indifferent to life or having failed continuously in its formation. Maybe, with only partial success. Or, on the other hand, why not, much more successful universes, living and dying at the top like healthy and rich celebrities. In conclusion, bearing no envy whatsoever, a set of different alternative histories. Along these lines, is there an indefinite number of other realities too? With an indefinite number of other theories about all these realities, with other theories about the multiverse itself? Is the multiverse a theory or a hypothesis? Is it a fact? Is it a better theory or hypothesis the one which states that every living being is a universe, a whole history, a reality, reality and history itself? Is the multiverse hypothesis contradictory in terms of this previous statement or is it essentially its source? But, is there a multiverse of multiverses too? A complex multiverse infinitely scaled, fractal? Are the relationships between living beings and the entire universe a fractal relationship? Meaning by that, is it the relationship between two scales of the same object? Then, for instance, if one is born and dies is the other also born and does it die in another way, in another scale? Or in other words, if a bacterium and a whale have parents, do the subatomic particle and the entire universe have some sort of parents too? So, is the multiverse hypothesis or theory the result of some umbilical selfishness that sees something strange outside obsessively, suspecting and distrusting its own shadow? Or is it a more than logical and sane reasoning that has multiple branches and a lot of other twin scientific concepts?

It is not difficult to find examples in classical fields of scientific study where the multiverse is perceived as a welcomed guest. In some of them, it is present through its own name or indirectly mentioned, but sometimes we also find it with a different meaning. Even, in some flexible reasoning contexts, it is presented as a new fact, almost magical, a kind of treasure discovered by science and interpreted as something really strange that can radically change the world, although nobody really knows when, where or how. As examples, by idealizing scientific approaches, it seems a concept that is linked with the neuroscience field that inquires and describes theories of choice, also with the most common and superficial interpretations of quantum mechanics – the Copenhagen interpretation, and so on. Thus, one can ponder or build precarious nexus between them; for instance, is there a multiverse, actual or potential, and one chooses to be in a universe or in another? Is this an irrevocable and daily fact, as obvious as the presence of light for the sighted being? Or perhaps, is any possible choice contained within a single flexible universe, within a single universe that already has all the roads ready and available for freedom? Does the multiverse diverge like a tree and its branches, like the heart and its arteries? And then, is each fruit a single universe, the result of proper elections or being shoved? Besides, is each outcome the seed of a new tree, of a new multiverse? But, does choosing help at all? Is there something that gratifies desires? Is there somebody with the relevant power to fulfil choices? Can the universe or the multiverse fulfil wishes as if it were a Wise King?

In the same sense, but speaking specifically about choices and the multiverse in a world governed or described by quantum mechanical laws, is the system of simultaneous superposition of states a good analogy to represent the multiverse? That is, is the multiverse a system of simultaneous states, of simultaneous universes, that ceases to be when one of them occurs according to the state or universe that is observed? Does it depend on what was chosen before and only after it is effectively observed? Is there anything or anybody merging election and observation simultaneously and indistinguishably? Otherwise, does one have a look at the programme schedule first and then watch a particular channel with the remote control, thus collapsing the wave function? Or as stated in another interpretation, is the multiverse always a multiverse but does the observer diverge, as if it were a common ancestor evolving in each of the parallel or decoherent universes of this multiverse? If this is so, would this mean that there is an observer entangled in all possible universes, in each one of the possible choices and observations that define the number of states of the system? Is the latter possible, is it possible to choose and observe all states and universes at the same time? A single observer who lives a thousand lives but only has time to devote to one of them? Perhaps, is there a multiverse of disconnected observers, a thousand egos who do not know the others, without any capacity or possibility to choose, living within the deterministic universe that the lottery handed out? Is there any difference between these and other interpretations, are they speaking about the same facts and with the same words or concepts? Is any interpretation of the quantum world a very common situation once translated into the world of personal and social relationships, of parts and sets? Then, for example, is an I a quantum? A quantum of what? A quantum of personality, society, us, memory, experience, knowledge, thought, universe, multiverse, information, code, tradition, freedom, creation, void or nothingness?

To argue more solidly on possibility and probability – especially in the field of biological life and its origin within this context of the multiverse – for a moment one must stick to, as far as possible and without creating a precedent, what is academically explained in the books of traditional biology. So, it is obviously assumed, as the God in the Old Testament, that life prospered enough to be a subject of observation and study; indeed, to the point of being now something so immeasurable that does not know itself, which now requires self-study, as if it were a traveller who returns to his village after wandering like a nomad around the world for a very long time. Therefore, in this traditional context, the miracle of the emergency of life is an issue removed from the fact of living today, which was necessary only in the past and perhaps it now requires impossible explanations; that is, explanations that involve giant numbers or numbers as incomprehensible as infinity itself, ineffable words or indefinable concepts, zeros to the left of the letters and to the left of limited reasoning. So, anyway, the common ancestor came forward, flourished and now there is a very diverse result; nowadays, life can be clearly observed, grown and folded, great inner-life able to reflect and consume itself, able to colonize much of the Earth and perhaps with the hope of conquering much more, of colonizing the entire universe or the multiverse too.

Then, whenever we speak about a universe among other unique, isolated, parallel, alternative, quantic universes or among universes connected by wormholes or agreed and bound by virtual or factual strings that form a network called a multiverse, or whenever we speak about only a single and complex universe without any actual multiverse, in any case, once the rebel observation is present and conscious, a certain perspective makes it impossible to describe this living universe as a place where everything went and came out without any difficulties, easily. The travelled history and the step of the current world – at least what has been narrated, told or written in some oral traditions, newspapers and textbooks – is not exactly or exclusively a lubricated gear, neither the smooth touch of silk nor a bed of petals with the description of its pungent and romantic scent. If this was a universe among many others but having won the lottery, a graceful universe, the other ungraceful universes would be terrifying, unimaginable, very unfortunate, killing life just for sneaking a look at its back involuntary. Here, apparently nothing has happened without conflict; because this universe is also an eventful history, harsh and grim, and maybe – pulling the elastic sensitivity of each one – also a very violent history, brutal and aggressive, without peace, a history written with seas of blood, sweat and tears. Obviously, what peace can there be in a universe born out of an explosion? Doesn't an apple tree bear apples and a rose bush roses? Also, what peace could there be in an asymmetrical doomed universe? What peace can come out from natural injustice? Is natural injustice the original and powerful gear of the world? Besides, is injustice the source of something more than revenge, greed, envy, and so on? Is a spermatozoid that gets first to the egg an injustice for the other spermatozoids? Is it possible to avoid injustice in all cases, in every action? Is there justice whilst living in any of the universes of the multiverse?

Let's assert the history of this universe as a history of natural violence but without linking unconditionally violence to drama — that is, to the subjective suffering or to the pain that one feels or takes pity on, like in the sting of a sharp needle on beloved fleshy buttocks — thus told, untied from any implicit order of non-violent nature, this is really a tremendous and tumultuous history, crass and crude. Since it is the history of gravity compressing mercilessly, molecular nebulae collapsing and forming radiant and ionizing stars, spectacular implosions of old and exhausted stars becoming overcrowded mass black holes that want more, such as gluttonous cemeteries. The history of massive stellar explosions capable of transforming the elementary, able to relativize the power of the largest atomic or hydrogen bomb built by human beings to much less than half the power developed by a weak hummingbird fart, far less powerful than the last gasp of the little insect eaten for breakfast when expiring deep in its mouth. Also the visible history of countless and timeless photons moving at top speed, to and fro like innocent particles until they are swallowed by the prison of infinite introversion of a black hole. Besides, entire galaxies colliding, stars and planets broken into trillions of pieces, atoms and molecules increasingly bouncing against each other, particles vibrating and spinning until they get dizzy and vomit numbers and other abstract things. In the meantime, protons attached by the strongest force against their will, erupting volcanoes and magma storms, meteorite and satellite impacts, fire tornadoes, thunder and lightning, ongoing floods of sulphuric acid and ammonia for millennia, earthquakes and extreme temperatures able to roast a whole cow al dente before realizing that the tip of its hairy tail is beginning to burn; chaos and darkness of all sorts. In short, far away from paradise, even for the most daring sadomasochist, for the most willing heart that would go voluntary to all the wars, to every disdained, scorned or unrequited love.

Then, if one wants to see it in this way, there is nothing more than this awesome cosmic spectacle, a violent and hostile world going towards an end; perhaps, to the indefinite expansion and therefore to the absolute final separation that will bring the rest through the most radical method, ending any relationships for good. But, is the universe constantly such a terrible place? If so, will it be in this way until the absolute freeze or crunch of the whole? Until the stones that have been angrily thrown can never achieve their target whatever the force used? Or with so much darkness, silently, a ray of light shines through and makes room, wiping out this reasoning? Maybe, something bright that has always remained alien, overcoming all adversity, overcoming any supernova or black hole effects from the very beginning of time? Or is it something new that emerges at a new moment in time? Is it the so-called life perhaps? Was life the first sign of peace in the universe, a healthy beginning that still exists? Is life a symptom of peace? Or quite the contrary, is it the supreme symptom of war, the absolute symbol and complete assertion of the actual conflictive nature of the universe? Is life peace or the central battle of the war? Isn't perfect peace a dead stone on the ground? What about a giant stone that flies around? Is the word *peace* a good choice to broach this subject, which other could be better? Anyway, is it possible to have peace in the question? Is there so little peace as in the answer?

Giving an opportunity to peace in the world of tiny probabilities, was peace or a better synonym the first and necessary condition for habitability? Was it an immediate ancestor of life, the very field where life flourished? That is, did life just emerge or bloom when the planet began to orbit rhythmically in some harmony, around a star of a certain age and physical conditions, at a certain distance, with a synchronized moon by its side shaped and calmed after big impacts, both in a very little troublesome zone, in a leafy neighbourhood of a specific galaxy far enough from other galaxies, large black holes and intense ionizing radiations, and so on? Was it only then, when it was no longer pouring with acid or running incandescent oceans of magma able to dissolve the stones before touching them, that life stuck its head out of the window? Then and only then, could something so pretentious, but so weak as a cell or a gene, take the liberty of leaving the burrow? Was that the very moment, when sense found some ink and made a word, when the prince found the foot where the glass shoe fitted, when nudity found a dress to wear at the party, when information was stored in a molecule and began ordering and dominating the environment? A natural consequence of a state of things? Is peace this state, the background or the framework of life and the opportunity to perpetuate? Did Christ give and leave this peace, the inhabitable world? If so, allowing the hostile will, deadly violence to befall him? Or did he simply know or understand life conditions, the need for peace, and then he practised and preached by example? Then, is this the peace that takes place when sacrificing an innocent lamb or eating an impala too? The habitability of a world by nature confronted, hungry and chaotic? The peace of a full belly, born from the overabundance of light and krill? However, was this exemplary and excellent behaviour a complete failure, an insufficient fact? Can perfection be insufficient, a droplet of perfection volatizing amid the sea of fire as if it were water, the liquid water needed for life? Perhaps, did the meteorite that impacted unwittingly on Earth in the Jurassic and wiped out the big dinosaurs carry more weight for the prosperity of mammals and singing little birds? Was this a work of a cosmic Christ or of a cosmic Roman Empire that was exterminating innocent dinosaurs? Nevertheless, did Christ finish the war and its causes? Or in the light of the issue, do the time and space of his peace span centuries mysteriously?

Undoubtedly it seems that violence, destruction and shocks, envy and hatred, greed and resentments, did not end with the emergence of life or with the modification or mutation of the past; on the contrary, it seems to become increasingly more evident and personal, more intimate and psychological over time. And today, carnivores still continue eating raw and roasted herbivores; and omnivores still eat herbivores, carnivores and other omnivores, sometimes with sauce but just to stay alive and occasionally they seem to have the consent of the same food too; besides, vegetarians kill or harm plants, the hungry ascetic hermit while sitting quietly kills fungi and bacteria just by breathing. No healthy living being would seem to have a motive against its own immune system or overloading it with work. Thus, any sacrifice or behaviour that avoids self-defence is understood in terms of altruism or generosity; that is to say, to defend, allow or promote other living beings that really fight and eat. But, is this violence natural and legitimate? What could explain the nature of violence, war within life itself, war between parts of the same tree? Two apples or pears spitting at each other while competing to death for the sab or millions of spermatozoids making their way in order to cross and be the first to get inside the egg? Why do trees allow their branches to be in conflict among themselves? Isn't there enough space to grow in an infinite universe? Is the tree of life only an empty concept, is life a true whole? If so, is it only a set of enemies that share the same code, the art of war? Do spermatozoa compete with sports rules, are they forbidden to elbow? Are natural laws the rules of their sport? Is the pursuit of victory and dominion the inevitable beginning of a war? Is war and fighting inevitable when one is an inefficient living being of an inefficient nature of an inefficient life? Why is limited efficiency not enough? When does overabundance end and when do lions and impalas start the race? In any case, if one fights, is everything a potential weapon, are any other living beings also an enemy too? Is a friend only an enemy of the enemy? Is a friend only a defeated enemy or a prisoner of war seeking a less humiliating and less painful custody?

In a new approach to fighting and war – aggression or violence, size and volume supported by the number of slave units, law of the jungle and methods of survival, competition and its zenith or cannibalism, psychological disorders that hurt oneself and the others, etcetera – in any of these conflictive patterns it is easily taken for granted that life is constantly in some sort of debt or is a burden, with pledges that must be inevitably settled, with knots or bonds based on more and more advanced or evolved contracts, indefinitely more complex than the original sin inherited from the common ancestor; indeed, the commitments of cells and complex structures of current living bodies. Again, there is the result of the confrontation based on the ideal Second Law of Thermodynamics, which makes it imperative to fight if one does not want to end like salt dissolved in the sea or like a slug flattened under the leg of a hippopotamus; if one does not want the body to finish like the cadaverous body of a cold-blooded reptile; meaning by that, being under the dictatorship of the Zeroth Law of Thermodynamics. So, everything seems in life to undergo a persistent and inevitable struggle until the end.

However, it seems also true that many battles of this war are faced in a different manner – one is a banker, a person who pulls strings, a modern Major-General or an unfortunate, orphan and sick soldier – thus, complete responsibility is transferred temporarily and part of the debt is postponed or guaranteed. So, the liquidation is delayed by means of working, paying in kind or getting on someone's back; but any deferral is in force only until death comes, then justice is done with the one in particular, however poor or powerful. Only then, the warm-blooded body is thermally balanced with the environment and the fight really ends. That is, natural peace between the body and the environment is installed, both at thermal level and at all other levels; hence, it all goes along the same way, there are no more personal, couple or social conflicts, there are no more diseases. Accordingly, once the world is calibrated in this way, is justice peace? Is death the same thing: peace, order, equilibrium and justice? And when definitive death lurks, is there permanent peace, perfect equilibrium, mathematical order, eternal rest? So then, does one have to inevitably wait for death to rest in peace or can one start to practise? But, is death of living beings the peace of an exploded universe that did not want this war? Or on the contrary, is death the peace of living beings treacherously generated by a lone, crazy and wicked universe which was looking to pick up a fight? Didn't the cell build or use a membrane wall to be calmer, separate from the outside while doing its own thing, to have more peace and quiet? If so, is war the origin, the precedent, the soil from which the flower grows? Therefore, is life making room, a cosy home amidst an eternal violent battle surrounding it, amidst a bloody nightmare that goes on during wakefulness?

Certainly, analogous to a cellular wall or a cell membrane, like the ant's nest or the rabbit burrow, between the walls of a house, surrounded by fire or inside a cave, human beings also rest for a while; they can stop thinking about the hungry lion waiting outside, and because of this, the world opens up as if it were a window inside their heart. Then, maybe they begin composing music, planning and predicting; in the meantime consciousness becomes more widespread and more acute, seeing things from a new and different perspective, seeing new things too. And sensing freedom, they can go deeper and deeper, until they do artistic, logical and mathematical abstractions as well; finding the ultimate beauty of a new language, perhaps waiting to identify physical realities to match. Also, interlaced with the aforementioned, no matter if it is sooner or later, curiosity is fully awake; hence, privileged human beings begin to philosophize, do science and begin to sort out the recorded natural and human history. One begins to observe and comprehend the space and the time, to learn about the matter structure and the night sky, the innumerable stars, now raising its head without being afraid that something could bite its feet. In this way, an old boy is perhaps overwhelmed with the immense starry sky, away from the tyranny of the only Sun, the hot sun that burns and blinds, today's sun. And being on board a revolutionary planet he or she becomes a cosmologist, and after this, an astronaut who travels to the Moon or to Mars, then human beings know that they can go beyond despite not needing it at all. Thus, once secure in their physical and psychological dwelling, one can move faster than light, remember and travel to the past; one can stop time too, by imagining and plotting one travels to the future. In fact, one can even leave time alone and realise that the beginning is the end, and then get out of the vehicle and meditate; meaning by that, to be totally oblivious of oneself without losing one's trivial memory, but deeply understanding everything one is, always a temporary limited piece of thought tied to a powerful thinking or sensing machine. So, without committing suicide, at least human beings can die and see what happens.

But, before dying or stopping thought, is peace what allows this ideal human secure home? Is peace made through war with lions and other wild beasts? That is, through death, reclusion or enslaving others? If so, once this fratricide war is over, could more than one survive and live in peace? Would there be peace in his head? Or rather, all the past universal wars only to end achieving a personal conflict that condensates them or focuses them? Perhaps, is the source of peace only the war against evil? Being the death of evil absolute peace? However, what is evilness? Is evilness that part of life that goes against life? Who can afford not to go against life? What flesh is respected by the hungry lion? Is this always summarized by saying that we are all sinners, evil by nature, bad to the bone? And thus, does life inevitably lead to sin, to evilness? Thus, is eating a sin, a caloric sweet the greatest sin of a diet? But, is a sinner only the one who is aware of it, the one who kills what he or she loves? Is this awareness also a fruit of peace? Is awareness a miracle? Is peace a miracle? Is it a miracle to escape from natural war, from struggle and the inevitable presence of the Second Law of Thermodynamics? Who is it that does not know the apparent cruel smile of the Buddha, who ignores the result of eating the apple or the banana of the tree of knowledge of good and evil offered by the tempting snake? Who ignores the fact that oneself is fully naked of innocence, of perfect goodness? Who doesn't feel the shame that forces one to cover one's genitals with fig leaves, to cover what is made to reproduce and share this hungry, violent and fratricidal world? Is ignorance bliss or rather is bliss the murderous determination which does not hesitate or repent? Is facing the knowledge of the inevitable and terrible facts of the world in which one is involved part of the art of living? Is health affording the ignorance of omnipresent illness? How does one reach Nirvana? How can one stop the struggle and suffering, how can one extinguish desire, how can one stop causing or suffering the consequences in a world that does not allow nor represent the unilateral will of the living being? How does one become a zero, how can one destroy the solid ego, the material illusion of the I, one's reality? How does one reach peace, the only peace possible that bears fruit through direct relationship? That is, how does one head towards death? How does one do justice, the only possible justice? With closed eyes and with indifference, as one going down a steep slide? Fighting against everything for a while, fighting while sliding? Carrying the cross? If so, who better than the owner of the body to know the palm of his hand or to know what his cross is? But, who is the owner of the body? Is it the one who moves the hands and opens the eyes or is it the one who makes the heart beat or the brain think? Nevertheless, who would want a cross as if it were a sweet?

The arms used for the horizontal timber, the legs used for the vertical piece, nails to nail each duality to one single cross: horizontal and vertical, past and future, infinite and finiteness, definite and indefinite, truth and false, good and evil, observer and landscape, wave and corpuscle, particle and field, order and choice, order and chaos, information and expression, word and meaning, origin and end, genotype and phenotype, health and disease, physical and mental, physical and metaphysical, reality and illusion, right eye and left eye, northern and southern brain hemisphere, body and environment, you and I, we and they, awareness and ignorance, and so on. Everything and everyone, together in the same place and time, a crucial space-time event; a symbol of symbols, where all the parts meet again and one can use the valuable bank note, where the boy whom King Salomon wanted to cut in half is still alive thanks to his mother. The space-time shared by life and death. So, are all living beings crucified to life by the arms and crucified to death by the feet or vice versa? But, isn't peace and war also a duality on the cross? Both of them also crucified to unity without any options? However, is duality what is mutually exclusive? Is there war if there is peace? Is there love if there is something else? For instance, is there love if there is hatred, jealousy, envy, possessiveness, power, dominion, competition, and so on? Anyway, is the cross, as a unit of past and future, as a unit of the Alpha and the Omega, the centre of history, the eternal present, that which borrows and consumes from overabundance? Is it overabundance itself? Is the cross the overabundant present or is it who hangs nailed on the cross instead? Both as a whole for all eternity? What is the name used to refer to the unity on the cross that results from joining or crucifying life and death, is it usually just called life? Is it usually just called love? Does one frequently hear about only sorrow and death? Is it called cell, *gell*, organism or simply body or universe? Is there a proper word for this indissoluble unity that includes all existence and all movement? Could one say a few words from the cross in order to express that? What would be right to say if one has an engaged audience comprising all the things of time? Maybe one of Christ's last seven words? Perhaps: "Father, forgive them, for they do not know what they are doing"? "I assure you; today you will be with me in paradise"? "Dear woman, here is your son; son here is your mother"? "I am thirsty"? "My God, my God, why have you abandoned me?"? "It is finished"? "Father, I entrust my spirit into your hands!"? And among these, which of them fits better here and now, perhaps is the word that is a question too? A true *bioquestion*? What else could be said, what words could be added? Maybe, as the gospels of Matthew and Mark said: "uttered a loud cry and breathed his last"? Is a cry also a word? What is the hypothetical meaning of a cry of a dying man crucified on a cross where all the roads meet? Is it a cry uttered by someone that is enduring the pain of a wound? Or is it like the cry let out by whom was lifting weight with all his might? A cry of a newborn baby or the cry of a hanging body who avoids falling to the very last? The cry of a universal orgasm, the *grand mort*? An all-in-one cry, the cry of emerging synergy, or perhaps the cry of a Big Bang? Is it still possible to hear it and listen to it now? Is it something like the background radiation of the universe? Anybody's cry instead? What were the words of the two crucified men on the right and left of Jesus? Do they both represent the cross unity somewhat better from a biological point of view?

_6_

In order to awaken the wilful machine love speaks.

ψ

LIKENESS,
love and domestication

Beyond the death and resurrection of each moment, emerging and shaping a life, a point of view often commented gracefully, sometimes derisively, it is the one that explains or states that *it is always the others who die*, at least definitely. Nailed to the cross, tied to the electric chair, shot in the battlefield or in a religious temple, killed in a road crash, run over by a cycle or a truck, choked on a piece of bread, sleeping in an operating theatre or in bed; ultimate death always happens before the very eyes of those who are watching. At most, if one is particular about this issue, it could be said that one happens to meet someone in the dying process, such as two soldiers seriously injured and abandoned on the ground conquered by the enemy, expecting no mercy in the last moments, both dying hopelessly while holding hands.

However, this apparent fact, the constant death of others, does not seem possible to be said or understood by any consciousness whatsoever. Among other essential things, some experience is also needed, as well as reaching an age having enough physical and mental health to perceive that the living surroundings have come to an end and one has survived and still survives; whether it is the death of the closest and similar surroundings, such as those of the same family or species, or the furthest death of those living beings of other less similar species. In any case, the life that is taking place is always surrounded by the omnipresence of alien death. But, is absolute death something totally alien? Is death just everything that one has before one's eyes during one's lifetime? All change that is observed? If so, is there any change in oneself too? Isn't death the change that allows one to feel alive, a breath of fresh air? Is it also death which allows growth? A big black hole that pulls the string with which everything is tied and from which nothing will come back, such as youth? In turn, being life and the energy that sustains it only a temporary resistance to the gravity of this disease? Nonetheless, is it merely curious teenagers and adults with some experience who can question and challenge, try to understand and to comprehend, what is known as life or death? Just those who have the opportunity, the maturity and mental and physical health in a given culture? Certainly, does a given culture make it impossible to do so? What happens when one who meditates is totally oblivious to himself or herself, what about his or her life and death?

Moreover, for whom are the tons of written papers that speak and deal with these facts boldly? For whom is the *Bible*, the apocryphal gospels, the *Tanakh* and the *Talmud*, the *Quran*, the *Bhagavad Gita*, the *Upanishad*, the *Tao Te Ching*, medieval mysticism, Greek philosophy, Sufism, modern and contemporary western philosophy or any other cultural or scientific texts dealing with it? Are these just words, ink on paper, black on white intended for human understanding but requiring certain minimum physical and psychological conditions? So, is there the same limitation for all books and oral tales and traditions? For any meaningful or meaningless expressive information of the past, present and future? Does it include genetic information too? Is it altogether only a code used and specifically transmitted between the few members of a little private club? Is there perhaps information for everyone without exception? Just potentially, only if one accepts or argues some kind of qualitative marginalization? Therefore, if one claims universality, how could one explain – as so often has been argued – billions of men and women who have already lived and died, maybe dying during the first day after birth or dying at any age due to any cause whatsoever but absolutely unaware of any message of grace or salvation, enlightenment, peace, material transcendence or love? What was their life and death like? Did they love much less or were they loved much worse? Did they lose something irretrievable? Did they hear it from other ways and means, something more personal or natural? A sort of message without language, a telling silence? In any case, did they feel a need to know or to realize, to become someone or something, to believe in anything, to read or hear about something special, to do something that they did not do or to stop doing something they were doing? Then, taking this lack to the extreme, would the word of the so-called God be only for God, for an omniscient God who understands himself? Is biological evolution heading towards divinity instead of complexity, with the help of scriptures and teachings, with the help of knowledge and technology? If that is the case, and although one may be left behind right at the bottom of the dark sea, will the last one reach God or will he or she die on the shore? Will this last one be the singular automaton?

But, since one is still wandering and wondering in time, what was the life of hominids like a million years ago, did they already have time to listen and reflect on time and death? What was it like only fifty thousand years ago and what is it like today across the planet? What is the plain reality of the fact? What will logic, writings and speeches, language and the communication demonstrations be like in a million years? What will these activities be like when the same time from the emergence of the common ancestor has been spent, in three or four thousand million years' time? Will the *Bible* or the *Mahabharata* still be useful, will they be useful if they are preserved bearing the same proper and common names of the first day when they were written? Will it be enough to translate and interpret them as it has been done in a way so many times? However, if there is someone who wants to preserve the intact original scriptures, how will this one go on when the inevitable change or biological evolution has transformed the lips of the mouth, the teeth, the tongue and the vocal cords until they prevent us from the expression of an archaic language? Will it be necessary to be an automaton with microphones in the place of ears and a loudspeaker as a mouth? Has biological evolution maybe almost ended, will the human brain weigh a ton some day? Anyway, is there any transcendental knowledge usable only for a section of the current and historical population? Is this knowledge and its broadcasting a form of segregation? Is this perhaps what life truly wants? Can something like this ever be universal? If so, how will the irreversibility of ignorance and death be overcome? In turn, what role do animals play in this business, what about the current and prehistoric plants and all other living beings that have disappeared since the emergence of the common ancestor? Are they lost in the dark and misfortune, at the same level as a stone? At the same level as a wild man with no contact with others, at the same level as a child born dead or the one who lives in a baby awareness for a couple of days? At the same level as someone who spends a lifetime, from the outset to the end, in a vegetative state plugged into a machine or working like a slave in a factory, as if it were a business-machine piece? Then, is all kind of experience and knowledge nothing more than the icing on a cake, maybe the most attractive and delicious part but also the most expendable to feed the hungry? Is it rather a completely dispensable part, inevitably expendable when trying to feed everyone? Does everyone eat from this cake, despite only eating the crumbs of the crumbs that are falling down on the ground?

As seen today in most of the human world, the death of other living beings of other taxonomic ranks is often immediately associated with other concepts instead of alien death itself; indeed, they are mostly ideas or thoughts about food – protein, fat, fibre, vitamins, etcetera – fabric or clothing, building material, science, medicine, money, discomfort, entertainment and games, pest or disease, risk, even culture or art. Namely, from this it can be theorized, with centralistic coherence, that when things become less similar, when wider sets are formed, constituted by elements increasingly dissimilar or just sharing the basic features that define the resulting set – as could be an ideal classification of human beings or the chimpanzee in the animal kingdom and then in the complete biological domain of the eukaryotic – whilst the set grows, it seems that empathy, compassion, complicity, care, and so on, also generally tends to gradually weaken or disappear. Until it reaches, for instance, the boundaries of absolute indifference towards the individual life of a bacterium or other almost invisible unicellular beings, despite the indispensable ecological and physical underpinnings that these living beings sustain all over the world, as the commensal or symbiotic relationships of the bacterial flora of the gut of human beings and chimps.

Furthermore, beyond the human being that is regarded as a centre of mass, in all other species something very similar also seems to happen; in many of them, perhaps ignoring the presence of others at a minimum level of detail. At least, just as a lion is apparently unaware of bacteria but is very much aware of the presence of the neck of the impala; equally, as the impala ignores bacteria but is aware of the claws and the cutting canines of the lion as if it were a diamond and gold treasure. In turn, what do bacteria know about others? What would life be like for someone who feels the loss of a single bacterium as the death of a pet, a neighbour or a family member of the same species? What would it be like if the death of any member of the same species is equally felt? In this case, would one do something more during his lifetime other than suffering and mourning, being the source of an immense vale of tears? At least, would this one have a clearer and sharper perception of life? But, is this equity really possible? What enables it and what prevents it from happening? Perhaps, does the overabundance of love for the most similar prevent it? On the other hand, is global and local news on television, the internet and newspaper trying to promote it? However, if one hopes for an equitable attitude, without a nucleus or a centre, would it be absolute indifference the only possible and reasonable form of relating to the environment, with the internet, television and the papers or streets full of death and suffering? The same indifference that is towards bacteria while being healthy or when there are no benefits to be had? The same indifference that is able to see all dissimilarity as part of an isolated and alien group? Or would this attitude reflect boundless love that cares for everything as if it were its own child? Would this still be more preposterous, who has done it to practise what one preaches? Can this one explain the good points about it? Anyway, is likeness just a physical link, a shared morphological feature? Or is it there more to it?

Beyond prokaryotes, on the threshold of recognition of the face in the mirror, more strangeness and indifference is still usually found. In the domain of the virus, the plasmid, the stone, the star, the subatomic particle or the electromagnetic field; in conclusion, in the inert or inanimate world, even in the theoretical world of physical laws and other abstractions – especially, in the conceptual world of new quantum mechanical laws – there is a resemblance that seems to have completely disappeared, remaining as if dead; colloquially speaking, there is no mouth where the hook of the heart can cling to, no handles remain to lift the jar of water and drink, no concept of common language is analogous to the precise reality that is glimpsed. So, is the definition of life based on some kind of likeness and finishes just at the edge of recognition of this likeness? That is, is there recognition beyond likeness? What or whom does the whole life look like to be able to define it completely? What is the proper analogy, who or what can be used to compare it with? Anyway, who cares about the consequences of stepping, drinking, eating and breathing the air filled with defenceless and innocent bacteria? Who or what is loved by a stone or is in love with a stone? Who is in love with a stone statue or with a three-dimensional hologram that responds to touch and words based on a charming software programme? Who loves a pure diamond the size of a watermelon or a gold bar as big as the beam of a cathedral? When was a star, no matter how great and necessary, something more than just a ball of fire? Does likeness fade, dissolve or lose itself completely in ultimate and definitive death? Is death absolute dissimilarity with life? Does death have nothing to do with life? To whom or what did Christ cease to resemble when he died on the cross after the cry he uttered? When can one spot dissimilarity? Meaning by that, a different exterior, the outline of one's own hand and the background? Does this outline exist or is it an illusion, Maya, a holographic projection, a snake on the rope, a horizon in a sphere, the form of a ghost and its castle under a single sheet? Is there something more than a sheet spread out with nothing underneath, a sheet with dynamic forms or patterns that is precisely the entire universe and everything in it? Is there only an infinitely spacious nucleus, a beating heart without a pump inside? Is all just the shell of a soul without an egg white and yolk? Is it just a mask that is capable of pulling all kinds of faces but without hiding behind any true emotion or feeling?

If one can settle, as an initial reference, the taxonomic range of species and specifically one of the current species, the tendency to gradually lose the biological resemblance around it can be superficially explained through some figurative literature – as it will be done a little later in order to carry on asking more questions. Thus, before going down this convenient path that approaches the increasingly broad base of the so-called life pyramid – microorganisms such as bacteria and archaea – some initial questions can be put forward here to sow this soil; for instance, is similarity, resemblance or likeness finally lost within the number? Is it lost within giant numbers, within the gradually infinite ocean without borders? Is indifference the largest infinity, infinity itself, a merciless zero without any compassion? Or rather, is infinity impotence, a powerless zero than can be no other than what it is? However, is it possible to be different from what it is? Can the verb *to be* mutate radically, resurrect newly created? Or does it have to be happy with conjugating itself or ceasing to be? Is the mesoscopic world lost within a too macroscopic world and within another too microscopic one? Is the cell the microscopic limit and the blue whale the macroscopic limit of life? Is a virus or a plasmid the limit below the feet and human beings the limit just above the crown, the trough and the crest of the same wave? And far beyond, is there one infinite below and one above, an inert and dissimilar infinity, subatomic particles and starry skies? Is humankind the centre of a linear world, the midpoint of a number line, the intersection of two infinite lines or any other functions? Is it rather the intersection of two figures or sets? Only two? Aren't there crosses with many more arms, perhaps some infinite sphere absolutely full of solid matter? Is humankind itself the intersection of all things that could be said, the cross of infinite multiplicity? Or is this the place for something smaller, something between the common ancestor and the largest dinosaur that had ever inhabited the earth, just like a freshwater fish or a tiny insect? Anyway, when retelling and recalculating space and time to their end, will the future living automaton be the singular centre of the universe? Will it be the new alpha and omega, the beginning and the end of history? Will this crucified automaton be the core of the world and all its passions, each bit of its body nailed to every physical and psychological dimension of a multidimensional universe? Will its heart be the conscious and aching heart of the multiverse? Or will its imagination be the tip of a spear while opening up the future, splitting and conquering nothingness, the leading or cutting edge of a living universe? Perhaps, after seeing its face in the mirror, will it refer to itself as the Tathagata too? That is, *that one who has thus come thus gone*, easy come easy go? Will it describe itself as that one unperturbed by change, upright and serene facing constant death? But, will it describe the face in the mirror or the face in front of the mirror, or in both places? Perhaps, just as Moses heard from within a bush that was on fire but didn't burn, a singular automaton would say someday as if it were an angel or a messenger of God: "I am who I am, the being of all beings, the living of all life, the death of every dead thing, what it is and what is not, *I Am* is my name"? Or will it say more reluctantly and simply: "I am also who I am, a son of God like all of you are"? Is this a basic statement that can be easily said both by a parrot and any human being?

But, if one goes on to question a little more about centres or nuclei, about terrible crosses or about joyful spatial-temporary events, is it possible, in any case, to escape from anthropocentrism, from the centrism of the observer who claims to represent part of the truth or to develop accurate science? Is anthropocentrism the axis of the most accurate science or is it the infinite limitation that really prevents observation from touching the truth, from seeing it alive for a moment? So, is it possible to draw a mouse without looking like *Mickey Mouse*, a mouse without looking like a human being, a human corpse without looking like a zombie or a corpse without seeming to have a breath of life? Which hand can do it? Is an artisan's hand enough? Is it possible to programme automatons to achieve it or will they also show automatisms? What theory does not require anyone to be formulated, communicated and verified? What hasn't been generated, stained or interpreted and finally spread by an anthropomorphic – or any other morphism – mouth and thought? Would this be a theory? Would this be an actuality? If so, how could one convey something about it? What has been made bearing no likeness with its maker? Is it possible to make a figure without clay, a form without information, a phenotype without a genotype? What message, what knowledge or science is absolutely independent and universal? Is independent and universal the silence stamped on white paper alone? Maybe, just the forms of the fire while the paper or the bush burns? The shapes of the clouds blown away? A fossil of an extinct living being of the Jurassic? The pareidolia that results in seeing simplistic faces on everything, on the Moon, on Mars, and appearing on any surface of light and shadows? What form should appear to satisfy instantly this demand for purity? Must it be a form? Moreover, a cognoscible form? The human word coming from a wild untaught parrot? The shape of a laughing and weeping face lined on a rock that emerges from a volcano? A *Mona Lisa* or the *Vitruvian Man* tattooed to detail on the back of a newborn dolphin without genetic engineering intercession? A cloud forming a word, writing it while breaking up in the sky, writing the word *cloud*? What font type would the wind use? And avoiding the inevitable hindrance of existing different languages, something like an intelligent rainbow of words appearing only in the language of the reader according to his own relative linguistic position? In any language, the full *War and Peace* written by the trail of a worm? A rebel worker ant that takes control of the ant's nest, subjugating the queen and releasing her to give birth for a while? A nebula resembling a finger, pointing to a galaxy full of star systems with intelligible life? A black hole that becomes a white flower spreading its luminous scent? The paraeidolia of a human face superimposed on the face of an ape? A crocodile imploring to be ridden or a fighting bull praying to have its neck thrust with a pair of banderillas? An impala putting its healthy and strong neck into the mouth of a lame lioness?

So now, whatever the exact place in the universe where living things settle and graze, a story of similarity and dissimilarity can be told, made out of broken pieces; in this case, based on and derived from an average human being, an ideal Homo sapiens certainly inexistent but that is useful to describe firstly, ideally, a current cousin of his; for example, another primate such as the chimpanzee. An animal usually with two eyes, a mouth and a tongue, two ears, a brain with neocortex inside its skull, two hands and two feet, five fingers at the end of each limb, face, a nose with two nostrils, two elbows and two knees, two nipples, penis and testicles or a vulva, neck, occasional bad temper, social meaning and a use of social skills, perhaps a sense of humour too, surely a sense of justice and proper morality – undoubtedly, clearly knowing right from wrong. In short, a cousin of the human being that would spend most of the overabundance of its time – that itself represents and is – on trees, while resting or hanging from the branches. Also, walking on the ground with the use of its arms and legs, sometimes in bipedal form like chickens; using rudimentary tools such as its teeth or a stick too. Apes that defend themselves and become strong in the highest places, obtaining some freedom there and attacking the others in some circumstances.

But, is it known whether chimpanzees have or apply any *jus in bello* or *jus ad bellum*, some sort of war laws? Do they sign treaties after a bloody day? Is the chimpanzee a fraternal cousin of the human being or, as in the case of the excited and relaxed electron in the quantic atom, isn't there anything between species? Isn't there even a lonely photon that represents a transitional fossil? Are there humanzees as there are nectarines and mules? If not, are they a possibility or are they an impossibility, what does it depend on? Among chimpanzees, are there those who are more or less similar to human beings? On which measuring method would it be based on? In this sense, what are the limits of genetic engineering? Is it possible to overcome natural limits? Indeed, is it possible to make up or invent something new that nature could never make up or invent, even devoting all the time of the world to evolution? Do chimpanzees laugh and cry? If so, is it evident when they do? Can one distinguish a chimp or a human being that roars with laughter from one who cries inconsolably? Can one at least do it without the context that helps to distinguish it? So, is the boundless pleasure of ecstasy and the sharp pain of the open wound the source of the same facial expressions? Do chimps have a morality that takes into account something that goes beyond them? Maybe, the goodness of the tree that yields them fruits? The wrongness of those heavily armed big cats that climb trees to invite them to dinner and insane party?

In order to continue along this path, after mentioning a great ape like the chimpanzee, with a DNA that is so close to the human DNA, some of the second cousins can be mentioned now, other terrestrial mammals such as elephants, cows, sheep, cats or dogs. Generally, with more than two nipples and usually walking on all fours, more inclined animals, more horizontal. Meaning by that, on the whole, less similar to a Homo sapiens than a chimpanzee but sharing the same basic features: two eyes on a face that is vertical to the ground, mouth, lungs, liver, heart, brain with neocortex, ears, teeth, neck; also expressing anger when they get annoyed, with young animals playing around their parents, a language and a more or less structured society, some rules to abide, and so on. Likewise, just in order to indicate some sort of superficial dissimilarity within the large group of terrestrial mammals, other animals with a little more transfigured morphology can be mentioned here too, such as bats, mammals sleeping upside down and flying by clenching and unclenching their fingers; or farumfers or hairless mole rats, underground burrowers, eusocial beings without hair, colour, skin and pain – offspring of a reproductive queen, like the queen of a swarm of bees but with far fewer subjects.

In addition, pressurized inside this closed and claustrophobic drawer, we can now mention some marine animals such as blue whales and striped dolphins, *milk cousins* but inhabitants of a radically different environment, in which they have been adapted by profoundly changing food habits, morphology, etcetera; lineages of a privileged intelligence, but without anything like a pair of adequate hands to make it effective and perceivable. Animals that on fleeing or on the search, immigrated from the solid earth to the vortexes of the flowing water of the sea, surrounded by liquid until they went bald; however, without settling down definitively since, as if they were nostalgic beings, they remain tied to the air, always plunging while remembering that they have to return. Thus, back to the surface they breathe through the nostrils or blowholes, opened at the top of their head, in order to dive again, waving the caudal fin as if it were two monkey legs welded to its tail, as the human swimmer does when practising butterfly.

However, is it the freaky elephant with pink ears or the circus elephant dressed in jeans sitting on its bottom and painting pictures with the trunk a proper element of these categorical sets of mammals? And the dolphin swimming and playing with a plastic ball in the aquarium waiting for an easy fish to catch? What about the sacred cows that roam the populated cities without observing zebra crossings or traffic lights? And some of the pet dogs that eat and live much better than the mightiest of kings, emperors or maharajas of ancient times? What are the living conditions of the queen of hairless mole rats? Are they very similar to those of their offspring? What would the world be like if the queen of the farumfers gave as many offspring as the queen bee, what should the planet be like to make this possible? What would happen if there were so many cows and whales as krill and blades of grass? Would a pack of lapdogs or cats the size of a large elephant behave as lovely pets? Who would be the pet then, who would be the tamer?

In any case, assuming that domestication is evolved tameness – the genetic effect of taming over time – then, are both, taming and domestication, processes that includes making similar what is dissimilar, not unlike what is unlike? Is a domesticated chimp or dolphin more like a human being? If so, domesticated to do what? Is it possible to teach what one cannot do? Are the fruits of a plant that is domesticated to be less toxic than the wild variety more like a human being and his spoilt pet? Is this domesticated plant perhaps more like the human taste and his blood serum, more like his healthier body or more like the sweet dreams of his immunological system? Which living beings are easily domesticated, what fundamental feature is characteristic of them? Is it just a DNA easy to handle, to make them still more docile and tameable? Is some kind of phenotypic likeness, at any level, also a prerequisite of taming and domestication? Is there some correlation, positive or negative, strong or weak, between similarity and taming or domestication? Perhaps, is a common language necessary whereby orders or instructions are understood? A more colloquial and interspecific kind of DNA, some sort of slang? A language that no one understood better than the dog? Was the wolf or dog the first tamed and domesticated animal or was it rather the first animal that moved closer and it was left to move closer, gradually shortening the distance? Are sheep, also one of the first domesticated animals, the most similar animal to a human being who domesticates it? Are sheep easily tamed for the same reason that the fruit fly is so much used in genetic engineering? Is this the same reason why woollen coats are made instead of barbed coats from a hedgehog? What are the limits of taming and domestication? Is it possible to tame a chimp, with the help of genetic engineering and a banana, until it is transformed into a human being? Or, at least, could it be transformed to have more resemblance to a human being? In turn, can a domesticated chimp tame or domesticate wild farumfers? What about wild human beings? Does the lion tame the impala in a brutal way? Is the natural or artificial selection a domestication process? Is it perhaps involuntary taming and domestication process, the result of a stubborn life that is always beating its head against the same wall, always going through the same meat grinder? Certainly, is making someone believe in something a process of taming? Can one tame oneself in this manner, would this have any meaning? Would this be a domestication process too? Is the teaching of baby mammals virtually the same as taming a dog? Was an atom, the water molecule or something similar, the first tamed and domesticated entity by the incipient common ancestor? Did it have to tame the simpler matter components or the natural laws first? Is it possible to tame laws or are laws the only true tamers, the tamers of chaos and darkness? Was the stone the first apprentice, the first believer, the first object tamed and domesticated by hominids from the Stone Age? Was the stone the first privileged object to know a better world? Is there any indomitable stone, a stone that does not want to be part of a house, a tool or a weapon? Perhaps, a stone the size of a planet, in motion and letting one know the untameable inertia of the wild?

Descending the face of relative resemblance – using a fragile rope and remembering that this is a path based on an arbitrary classification that is summarized and structured clannishly with little scientific rigour – the oviparous set could be put now inside a broken basket too. And firstly, reptiles, amphibians and birds can be named among them; for example, a snake, a crocodile, a chameleon, a toad, a goose or a talking parrot. These animals, again only generally, present a relative dissimilar structure; certainly, even more bizarre, transfigured and varied in relation to the prototypical likeness of mammals such as human beings, chimps or cows; with noses that are beaks, arms or legs that are wings, hairs that are feathers of all kinds of gleaming colours, foreheads that are combs, eyes that move irrespective of each other, yellow and venomous skin, backs covered with green scales; with veins and arteries carrying cold, warm and hot blood; with flat and elongated bodies or almost round like a ball, or bodies that lost their limbs to crawl better and which are nowadays easily mistaken for a rope. Also, walking bipedal such as human beings or on all fours as cows or cats; with horns, forked tongues and extruding and folding appendages. Land, sea or air lovers; monogamous or polygamous living beings able to swim like fish, to fly from one corner to the other end of the planet from months on end over the mountains, able to bite with the might of a metal sheet industry stapler or to climb smooth, vertical and high walls; or crawling on the ground and successfully tempting those who were living in paradise. Besides, a host of other features and other locomotion skills that makes them very difficult to be imitated by the elephant working in a circus or by a cow delayed due to the disrespectful city traffic.

Having said that, beyond the relative difficulty of understanding and manipulating their particular genomes, are these animals more difficult to tame and domesticate than sheep? Didn't they approach human beings as the wolf or dog did, following the scent of food heated in the fire? Are they overconfident in their approach or are they too afraid? Can one ride quietly on the saddled backs of a crocodile? Is it more difficult to saddle it than to ride on it? Or simply, does one take longer to know their weaknesses, how to tempt them to obey? Don't these species live by bread alone? Do most of them have an aversion to being puppets? Is it the inertia of millions of years doing what they like? Or do they lack or do they have some sort of genotype in excess? On the other hand, could it be that a crocodile is of little use in domestic tasks and therefore the art of its dressage has not been sufficiently practised? Is a crocodile inefficient to pull a cart full of coal like a horse? Isn't a parrot efficient to teach mother tongues? If it is so, doesn't a parrot speak like a three-year-old human being who recites a passage from a philosophical text or a love poem? In the same way that an eminent mathematician says that two plus two is four, four by four sixteen, the natural logarithm of one is zero? Like the better physician who writes and solves differential equations relating to mass and energy?

By making a virtual chimp step more towards dissimilarity, still among the oviparous, a big holed net can be used to encompass most fish, such as the whale shark, the swordfish, a sole or a scorpion fish; stated here as being less similar to human beings or chimps than dogs or parrots, if only because they are water animals that breathe through gills, without limbs, hair, feathers or any fingers. They still have a skull and a spine, but with more thorny ribs. Animals with fusiform bodies and cold-blooded, tails that become confused with the trunk while integrating stealthily with the face. Certainly, with diverse bodies; flat and seabed sediment coloured or with bony and elongated faces like a bullfighter's sword prepared to go in to kill a liquid bull; with hairless homogeneous surfaces, sometimes surrounded by poisonous fins as if it were a sort of sun and its rays of light, with brightly coloured scaly or grey skins, with yellowish freckles looking like a checkerboard; with eyes on the sides of the head, small like buttons on a coat, or two eyes at the top of the head, both very close and resembling a periscope watching from underground. Also, hearing by means of multichannel holes without ears, eating with mouths that hold vestigial teeth and that would swallow a cow accidentally, or with rows of teeth in the shape of a saw. One could say much more but, how does one know if a fish cries? What does one need to tame the scorpion fish as one who tames a dog or a sheep? A smaller aquarium than the one built to tame the whale shark, an ocean in a drop? Are harder methods needed? Some pheromones and a big maul waiting for the opportune moment? Is death a method of taming and domestication? A method for taming species and their behaviour, the ideal tool for artificial selection? Is final and definitive death the final and definitive taming of life? Is definitive death the domestication of permanent resurrection, the end of the changing and evolving life? Anyway, is evolution towards complexity evolution towards dissimilarity? Or rather, towards becoming dissimilar, moving far away from the common ancestor?

A breath less similar than most fishes and being only distant blood cousins of mammals, let us mention now terrestrial arthropods such as insects and arachnids; another huge group, the most diverse of all oviparous, which is categorized within the wider animal phylum on Earth, the group of invertebrates, animals without a spine. The anonymous mass that is already beginning to blur in the so-called dissimilarity. Animals that have segmented bodies and are counted as stars are counted, billion by billion, without using fingers. Lovers and messengers of plants and flowers; and also, heartless carnivores plucking heads with a single jaw movement. Insects with compound eyes as large as the rest of their head, with two pairs of wings and six legs around the chest, such as the bee, the mosquito, the fly or the ant queen. Or poisonous arachnids with eight thin long legs as the black widow, a sexual cannibal; or the scorpion, a viviparous with two clamps and a sharp sting at the end of a segmented tail. All of them with the skeleton on the outside, as if it were still the shell that protects the egg yolk, an evolved shell; also, for instance, with protuberant horns and trumps coming from relatively small sized bodies, to the point of being as small as the length of the body of the parasitic wasp, that is, as long as the depth of a sheet of paper.

But, do the spider and the insect belong to the world of indifference? Can one love an insect or a spider? Is it the same to love a particular insect than love insects? Is protecting them to love them, wanting to eat their honey and all the trees of the garden be pollinated, to harvest their sweet and slightly toxic fruits? Is loving them wanting them to grow healthy and strong, also wanting to feed the birds that make the sky and the earth a musical place? Then, is this love based on ecology, loving only in terms of covering physical needs, due to the private benefits obtained, whether mutually or unilaterally obtained? Is love the bond or the link of the members of an ecosystem? Is the relationship between the impala and the lion a love story, a passion game played until the end? Do people love their neighbour, partner, friend or parent for something specific, because it satisfies a physical or psychological need? Indeed, as part of a personal ecosystem? Is it because what is needed is offered and vice versa, such as providing drugs to the addict or receiving drugs from the dealer? Or does one love and is loved because of likeness? Because of the likeness of hearts? Is resemblance or likeness necessary to live, necessary to love? Is loving to realize that another is similar and then act accordingly? Is love a conscious or voluntary act? Could love be conscious and voluntary without being denigrated in the theatre of the mind or stained with the hands? If one loves dissimilarity, does one then know what or whom one loves? If so, how does one know, can one understand or comprehend what is not? Maybe, is this what is usually loved, only what one is not? Only because of being indifferent or rejecting what one really is? Only because of loving beyond oneself, due to the overabundance of love? Is it possible to love a mosquito that sucks the blood with an infected trunk without permission, at night and treacherously? Does one get some kind of hidden benefits from it along with a slightly annoying infection or the injection of the parasite that causes malaria? Does one perhaps learn robotics too, how a biomimetic mosquito robot must suck and blow when attacking targets by injecting nuclear nano-bombs at night and treacherously? Are mosquitoes necessary in order to feed birds or continue selling insecticides? What does the mosquito think about this last question, what does it answer? Is it possible to tame and domesticate a mosquito, a bee or a black widow? Is beekeeping an example or is it like having a bird in a cage or a dolphin in an aquarium? Is a cage or a jail a warm and comfortable house full of food where some pet dogs live in winter? Is love a cage, a jail full of prisoners? Is the farm a lifetime imprisonment for sheep, always waiting for the last minutes in the slaughterhouse? Is the farm worse than the slaughterhouse? Is any form of taming or domestication putting living beings in a cage? Something like an invisible sheet that slowly, like a cloud or fog, covers the other completely? A sheet or a dress made with knowledge, intelligence and strength to change and take advantage of others?

In the same way, is it first necessary to tame the other to be able to love it afterwards? On the other hand, does one have to love first to be able to tame? Is love what one feels for a domestic spider that has been named and is fed on flies and mosquitoes? Is love a feeling or an emotion? Are sentimental and emotional people those who love more and better? Does love have any kind of gradations? Is there partial or limited love? Is a spider web a declaration of love? A way of saying *I love you come with me*? Is love being hungry, the desire that one has to eat and to do something with the energy provided by food? Is love for food and drink a love as strong as a stone rock? Aren't the eater and the drinker similar to food and drink, is it because of this that there is love? Do carbon and water love carbon and water respectively, pressing to form a unity? Like loves like? As the proverb states, *birds of a feather flock together*? Are only the carbon and the water of the lion pressing to form this unity? What about the carbon and the water of the impala or the grass that it eats? What about the carbon in the stone on the ground or the water flowing like a river into the sea? Anyway, is all form of life also food and drink? Who, in the tree of life, is exempt from being food for another? Are exempt of this those who for a moment of glory are at the roof of the trophic food chain or in a privileged node of the trophic food net? Is the top of the trophic chain or the privileged node in the food net a static ideal? Is this quite a dynamic throne of a spinning wheel that never stops? Is this, again, the big wheel of Samsara? Are human efforts the vain hope to break this wheel, to climb to the top or to go into the fortified and satiated node that is surrounded by the world of hunger? Does one hope not to end up being also a steak or a hamburger, a milkshake or a sauce? Is this fact present and evident nowadays? That is, oneself identifying with food? At least, is it as present as food is served at the table? Is one more aware of this fact when missing food, when the body begins to consume its own fat while slimming down? Will the fractal universe finally eat up all life in one mouthful too, integrating living beings within its cosmic body and its universal laws in an absolute communion? And ultimately, is the black hole of black holes of death satiated? Is it enough for it to suck up to the last drop of carbon and water with a straw, until the last memory is forgotten? Perhaps, is it possible to fly away from gravity, to leave behind the wheel of Samsara, to let Nirvana arise? If so, is it completely necessary to have a full stomach and an empty mind?

Let's move forward, in order to finish with the rest quickly, swallowing the sea with a little human mouth and compressing the richness and diversity unfairly in a small space; then, one can name some more animals that are presented here, to continue the sequential description started, as being more dissimilar. Thus, marine arthropods can be cited, just as numerous; crustaceans such as krill, shrimps, crabs, barnacles or lobsters. Besides, there is the enormous and diverse phylum of molluscs, such as octopuses and squids, cephalopods with ink sacks, with smart brains and tentacles, some of them giant animals with three hearts. In the same phylum, gastropods such as snails, with tentacles with eyes at each end and one single huge flat foot dragging while simultaneously supporting their spirally coiled shell. There are also snails without shells, destitute as slugs, a head, tentacles, eyes and a foot that is a tail too. Finally, let us stop the animal world at some point, with cnidarians like jellyfish, gelatinous brainless sacks with filamentous irritating tentacles; or corals, small cylindrical bags with a mouth and tentacles, grouped in colonies, compacting and keeping their past by secreting calcium carbonate until they look like marvellous underwater stone cities.

After that, the kingdoms of plants and algae, which some describe as being soulless, without responding by means of screams or running away from axes and fires, standing in the space of a moment like the most valiant and reckless of warriors; but sometimes too, silently and fortunately, dominating time as long-standing empires, with slow movements that conquer mountain ranges; big forests with ancient trees aged thousands of years, still alive after a thousand battles, still noble. Vast and self-sufficient kingdoms, with lots of green forms, leaves and colourful flowers, profound lovers of light. In any case, a more detailed description is not required as well as being impossible; from a baobab or a redwood, through the grass meadow, the fern, the rhododendron or a blade of grass in the steppe to the moss that seems a pillow of the rock or the unicellular phytoplankton flowing into the sea. And here also, suddenly and briefly, the fungi kingdom, a ubiquitous kingdom that is not necessary to dramatize a lot to make it a bit dissimilar, from the mould of blue cheese or rotten fruit to the yeast that ferments the sugar in grapes to make wine or to make the baked bread soft. Cryptic organizational structures that relate and mingle with everyone, with those dead or alive; bearing visible fruits as the white mushroom or the fly agaric, fleshy t-forms near the ground.

Finally, the incalculable number of bacteria and archaea, single or colonial beings but always unicellular, almost as simple as life is simple, almost as much as the last or the first universal common ancestor was. Similar to the basic structure, the building blocks that makes up the bodies of chimpanzees, fishes, insects or baobabs; certainly, entities more like the microscopic parts of these living beings, more similar to their cells than to the resulting massive set. Consequently, more like the parts dominated and forced to row in the same direction or rather, perhaps more similar to the parts that have freely decided to collaborate by doing yoga. In short, more like the body stripped from macroscopic beauty and name; that is, like the set of cells that are replaced from time to time, which are divided until the degradation of the telomere, until they grow old and die of old age; therefore, the living being which they constitute comes to an end too. But, meanwhile, is the soul this thread between old and new cells? Is it the form of the synergy emerged from association, a form sustained over time? Is it the name of continuity in a quantum world, in a quantum body, the bridge for the walk of life? Or is the soul similar to body cells which are not replaced? To those cells that are almost permanent, such as some neuronal cells of the cortex of the brain? But, is the soul just this, what represents what is resistant to the passing of time, alien to constant death, the childhood friend who is still recognized in old age? Is it the dynamic connection between two absolute strange eyes that form a single way of looking? Or the hope that one has, for example, when naming the newborn being, the parental ambition to call him or her always by the same name? Expecting him or her to be always the same? Is that possible?

Leaving the soul within the question, or rather, leaving the soul as the form of the question, also leaving viruses and plasmids out of life, a bit beyond absolute resemblance to it, and also leaving out the clusters of organic molecules that spend their time out of cells – too easily related to those inorganic molecules that instead of a cell form a stone – in short, once the biological similarities have been seemingly completed and finished, does one already know what the essential similarity is that allows theories to include something as a member of the sect of life? Does anybody know what likeness truly is and what things are undeniably similar among them? Is likeness the manifestation of the understanding between unique and rebel parts, between always new and free parts or individuals? Is it the love towards each other until they become totally confused? Is it the soul of the set, the spirit of the people? Is love what is poured between the gaps of difference and division, a universal solvent? Are beings and things a whole when tamed and domesticated by the same common domain, both by love or by hateful tyranny? By the love or by the tyranny of the observer himself, the one who compares and interprets similarities and differences? Is it everything that the powerful heart of a lover lights up similar? Is likeness the light reflected in the mirror of love? Is love, like gravity, an interaction that tends to join both similar and dissimilar masses? To unite all protons, neutrons and electrons under the same sky, under the same ghost sheet? Is it the law, even more essential than gravity, which integrates all kind of matter and energy to form a whole universe? Is it the physical law of everything which maintains a known universe constantly changing and still compound, still perishable? If not, does this law have a proper name? Does this name have a body? Is it just a force, the fundamental interaction that is simultaneously gravity, the strong force, the electroweak force and any other force that is or will be observed and described in the future? What law prevents a monkey, an elephant, an insect or a stone, from transforming into a person? What is a person, what is a human being? Is a person a stone with a proper name and being madly loved? Is a human being the one who does not love? Do stones love as well as being occasionally loved? Do diamonds and gold love as well as being usually loved? Does an insect love? Does the automaton love or will it love tomorrow? When is likeness of love lost, when one stops recognizing it in oneself or in others? Is love just some sort of behaviour or a feeling, something born out of genes, a mechanical characteristic structure that transforms over time? Can love be a peculiar programme or its relationships? Is it just a phenotype, an outcome, the result of a selection? Is it the extended phenotype par excellence? Otherwise, is it an extended genotype? An external order more strongly inculcated and powerful than the selfish genes themselves? Expressed and printed in the heart, stronger and deeper than natural laws, especially deeper than the Second Law of Thermodynamics and the Buddhist sentence? Something indecomposable and brighter than the light of a star? Something that cannot be absorbed because there is no power able to do so? Is love the tamer of black holes? Something faster than light, more basic than death, stronger than the final and definitive death? The cradle of existence that parents prepare for their children? The life that they disseminate into their eyes and inside a heart without substance, brought up from nothingness, from the zero that one knows naturally? The life that becomes similar to their love, born out of their love and extended part of their love? A new life resuscitated from the incessant dying relationship, a relationship incarnate and brought up again? Is love the wings of the shoes of resurrection? So, is it the source of creation too?

Once we have dealt with dissimilarities, but continuing with the narration that will come to an end soon too, let us return to the beginning so that we can discuss the origin of life again. But now, also talking about likeness, taming or domestication and love, a cocktail of common words that could probably be made with other words, and surely it is done in this way in another universe of the multiverse. In any case, one talks and wonders in order not to be still like a stone on the ground, in a universe where jobs are limited and controlled to preserve wealth and good things concentrated while poverty and bad things are widely shared around; that is to say, one talks to avoid paralysis, valiantly or rashly, rather than being silent.

So, in order to start the ending, let's state once more an episode of the *Book of Genesis*; that is, once translated into the language of the narrative that one is now reading, a part of the Book of Creation. A text that a lot of people from so many cultures and ages, let alone all of them, have read or heard about; therefore, some of their mechanical or repetitive descendants still know if they have not forgotten about it. For this reason, it seems to be more or less appropriate to rely on it and to say perhaps something else now. Thus, in the first chapter of this book, on the sixth day of creation after paradise was created, God says – although it depends on the specific text read – something like this: "Let us make man, a human being, a person, in our image, after our likeness, to be like ourselves". But, is image and likeness the same thing? What is the image or the likeness of the God of Genesis? If one wants to know about it, is it enough to observe, as the same *Genesis* said, the human beings or the people created? What could be better if one does not see the authors directly than looking at the work made in their image after their likeness? Who or what is a human being to infer what this God is like? Is it a plurality that speaks in the first person plural too: "Let us", "ourselves"? Nevertheless, is a human being a plurality or is he only expected to be so? Or on the other hand, is this God speaking to another character? Then, what other God is not named directly? Are there many other unnamed Gods, is it a divine community? So, is the spokesman of an infinite number of Gods speaking? The spokesman of infinity itself in all possible senses instead? No matter the number, is *image* all what people are, the expression of the absolute diversity or uniqueness of each one? Is *likeness* what they have in common, those things in which personal living beings are equal, shared things between different living beings? For example, as an ideal father and mother that recombining their particular DNA make up together the single DNA of their child? Hence, a child made in their image and according to their likeness? Meaning by that, according to what the father and mother put together, each of them a divine chain of acid, the two parts of the symbol? Is love the expression of the relationship of likeness? Is the love they profess to each other that in which they are similar? Is loving each other their likeness? Is this mutual love the image in the mirror where the child or the likeness looks at itself? Is the individual human being the expression of this divine love? Or rather, is humanity the expression of this kind of love, the trait of this divine genotype? However, when one observes oneself and the current human being as well those from the past, when interpreting the whole history, what is the image that best fits them? Isn't it too much responsibility, by means of likeness, to make anything or anybody out of God? So, doesn't the God of Genesis leave their nature in the hands of curious human beings, in the hands of erudite and generals, charlatans and soldiers? Thus, making what one wants or believes from God? Being God what human beings want or what they cannot stop thinking that God is? Then, is God perhaps some consciousness, an observer or a life that requires severe conditions to exist? One, two, three, twenty people or the whole material and eternal universe? Being God responsible for the struggle and the suffering, for care and peace? Being God overabundance and which consumes it to its privation and death, death itself?

In any case, does any description of God matter at all? Is the description the thing described, is the map the territory, and is the symbolic cross the cross of each actual personal moment? Anyway, does the concept of God try to lead the human mind, as if it were a shot of pure drug, towards eternal infinity? Hence, out of the brain too? Can words or concepts do that, can a brain do that? Which people can listen or be attentive at all? What can be expressed and reflected with words, words that the wind blows away like dust? Does any theology have any significance and value beyond the closed circle of people who talk to each other? Is it possible to be right in a way when talking about God? That is, what theological final exam would not be graded by the divine court with a very big zero, marking it with a perfect circle? Would it be enough to have all the time in the world, as in the infinite monkey theorem, to say something meaningful about God? Would this court give some points if those examined people wrote their names at the top and handed in the self-marked exam humbly: an absolute zero? Maybe, would the exam then be returned marked with a new grade: a zero, a comma followed by an infinite number of zeros and finally a one, which is a zero rounded up to one out of mercy and compassion? Besides, is God interested in God? Is God interested in looking at his eyes in the mirror, who would dare to put it in front of his face? Is the human theoretical field of study of God a reflection of God's thought? What does it matter to God what human beings said, say or will say about God's nature? Doesn't God know personally who is responsible for all things? Doesn't God know who God is, what God does, says, thinks, and what lies behind everything? Which human beings could teach or show God on opening their mouth, something more than vanity? How to make bread from wheat, how to make automatons from dust? Maybe, does God know human beings on listening when they speak about God? But, doesn't God know everything without having to listen to anything or anybody? On the whole, are all human thoughts about God a desperate attempt to kill or to tame God, to cage God in this world, within a confused brain? Therefore, attempting to incarnate God in oneself? Attempting to be omnipotent or hoping to arrange the neuron cells orderly as if by magic?

And simultaneously, beyond what human beings can see, hear or read, believe or reject, can God reveal to them or be a revelation in itself? What else apart from images of a prophetic vision or words of an inspired innocent mind? What else apart from the miracle of a universe bearing life or a human being who works miracles while asserting its own divinity and resurrecting to reassert it? What else can be said, what else could be the grounds for a dialogue or simply something to listen to? Are there limits within this hypothetical relationship between human beings and God or does the imagination run wild? Is everything imaginable also possible? Is this relationship like the parts want it to be? Does it take two to tango? Are there parts or any relationship whatsoever? Could humanity or human beings be related to someone more similar to themselves than they themselves? Can the eyes see themselves without a mirror or in a recording, see themselves in the current reality? Will everybody die without ever seeing their eyes from the outside and alive? Thus, is it only the others who have seen their look and know what they look like? Is it just the others who have the pleasure or the displeasure for a moment? So, can someone know oneself? Only through the eyes of others since only others see someone as a whole? However, isn't it always the case that others use prejudice glasses to look at others? So, in any case, is it impossible to know what one is like? At best, one can see only a part of his body, such his palm or feet? Then, is the self who wants to know oneself, just as one who wants to see his nape, always in a stupid catch 22 game, like a cat chasing its tail? Is there anybody without glasses enabling one to look at others clearly and therefore making it possible to say what the other is really like? By the way, is it possible for human beings to interact with someone else as a chimpanzee does with human beings? At least, just as a parrot says four words? Is the human being at the top of something and occasionally goes down for a walk? Or are they rather at the bottom of a hole? Can human beings relate to a being that has a higher encephalization coefficient? Would it be for the latter like trying to teach algebra to a beetle or playing the piano for the ears of a stone? Who is the human being with the highest encephalization coefficient across the globe related to? To the most approachable God? What is the source from which the most powerful brain gets inspiration and its ideas? Is it from its own brain? Does this source have anything to do with intelligence or other measurable and therefore limited capacity? Is the human being a creator? Someone actually made in the image of God, a son of God? Or is the existence of God the result of human invention, being God a child of humankind? If so, is God a natural creation that happens within the evolving universe? Is the universe an everlasting creator, is it creation itself? Is it an unconscious creator or an unconscious creation? So, what is the origin of the universe in terms of creation, who is its creator? Did the universe create itself or is it a creator without a creator? Is it simply the eternal constant creation? Is creation the untamed and undomesticated personality of the universe?

Still in the lines of the *Book of Genesis*, after human creation is recounted, the human being is crowned by God as the lord and master of beasts, animals and plants; the king on Earth who is set to kill and eat them without realizing this natural, thermodynamic and necessary evilness. Today, in the age of information, atomic energy and genetic engineering, it could be said, without a shadow of a doubt, also the lord and master of fungi and algae, of all bacteria and archaea, of the entire planet with all life that inhabits it; lord and master of part of the Solar System, the Moon, Mars, and time after time, step by step, perhaps in the new edited and published Genesis in the distant future, God knows in what format, he will be also called lord of the automaton and master of its mechanical consciousness too. Or surely, the living and fighting consciousness of the human being of tomorrow will be the singular automaton itself, and the God of the Genesis in the future will crown it as the new king of the known multiverse; the one with the power to choose a name for all dissimilar beings, making a universe of the multiverse through its choice, inventing its own personal world, bringing up its own creation.

Thus, in line with this speculative context, will human beings or humankind be the God of the Book of Genesis in the future? And then, will it be able to do more than humanizing? Is this a step below a God who divinizes? Is it part of the divine creation to humanize the universe? Does humanizing only entail growing and multiplying human bodies, human structures? Is humanizing also loving dissimilarity, from the living chimp to the dead stone? Or is humanizing making the others love dissimilarity? Can a stone be convinced to love the lion or the impala? Can impalas and lions be convinced to love each other too? Is this part of the actual human tameness and domestication endeavour? But, is it possible to humanize anything at all? On the contrary, can human beings lose their humanity? Perhaps, can it be lost if they cease to create, if they cease to love? Is humanizing linking nothingness with human beings or linking nothingness with God? Is humanity the transfigured image of God? Or rather, is God humanity itself, humanity that exists only as an adjective and an objective, such as the carrot is for the donkey? Anyway, one answers with arrogance the exam and shows to God a little vanity: is God the source of love that flows between dissimilarity and through it, as if it were a flood of fire that illuminates and warms until it melts down divisions and differences, until they are volatilized? A fountain from which human beings drink or stop drinking, give or prevent others from drinking? The love that eliminates distance and time, which opens the cuffs and minds to let hands and glances shake and touch affectionately? The love that generates more love? The love that allows humankind to defend its dignity without doing evil? So, the love that defends freedom, the absolute dignity of free people? Likewise is the crocodile defending itself from being ridden? But, is it possible to defend others and defend oneself at the same time? Isn't there always a final choice between one and the other, in a faced world that is naturally asymmetric, made of lions and impalas? Are people forced to be less than a perfect God that loves everything? Is this, precisely, the impossible perfection? Is this what biological evolution is looking for? That is, is life evolving towards humanity? Meaning by that, towards an actual universal being, a human being without needs and therefore made of the whole universe? But, anyway, is evolution necessary when there is love? But, is there a need of God if there is love?

And, in the penultimate sigh, before definitely ending this narrative, one can formulate some other questions, no more or less significant; for instance, how is a chimpanzee, a cat or a dog tamed or domesticated? What do human beings do to animals that would give an idea of what a God would do, did, is doing or will do to human beings? What is it that the best coaches do? That is, what are those doing who are loved by animals themselves as if they were a parent or a friend? What method do they follow? Do they free the animals and not expect any affection or work in return? Do they dress them up as if they were a person and expect a profit from a circus? Do they jail them in invisible, comfortable and safe cages? Or perhaps in a dangerous conflictive prison while showing them the key from the outside? What is the spirit of humanity to better understand what a tamer of human beings would face? Is it like the spirit of the sheep, the dog or the crocodile? In short, how to tame and domesticate a human being, what is the best and worst method? Is there a method or using a method is obviously the worst way? Can methods, systems and other traditions resist the passing of time, entropy, decomposition, and so on? What do the fathers, mothers and other educators or coaches of this day and age and of history say about that? What have the results of their taming methods been until now? Peace in the world? That is, peace inside the skulls?

Firstly, assuming that human beings are absolutely submissive and helpless, while the wildest and most powerful and deceitful of all the known creatures of the universe, is it enough to tame them by feeding their hungry mouths with a spoon? Is it better to lock them inside a cage made of gold and diamonds? Or is it better to instill fear with a stick and a supersonic whip and terrifying them? Maybe, inventing a forthcoming death lurking round the corner impatiently? If they do not fall to their knees, whip them more and more until the beast is tamed? Is it better to have a bit of music and some sport, some kind of entertainment? Action or romantic movies, gutter press or pornography, violence on television? Drugs and a crazy party, tourism and casual sex? A job, a successful marriage and a family to feed as if one were the food itself, as if it were the milk from the breast itself? Is it better to invent a God whom to pray for wishes, a Super-Wise King? In any case, what can one do if the frightened and poor beast does not come, if when drowning it still bites the hand that rescues it? What can one do if being submissive to the point of being depressed the God project does not start to bloom and starts to fade? What can one do if tangled in thorns, amid increasing pain, it is stretching and stretching without judgment or intelligence so that it cannot untangle the complex web that embraces it? How to tame those who hurt themselves increasingly and suffer deeply? Would one have to let them suffer until they learn, until they learn to die? However, is it possible through suffering to learn more than suffer? Has the immense past suffering been useful for humanity? If so, during the current human lifetime, is it better first to wound them with a spear, then insert a few pairs of banderillas and make them dizzy with a red cape? And once taught how to bow and touch the sand on the ground with their tongue, ending with a precise and vertical thrust between the shoulder blades in order to reach and traverse the beating heart? Would it be better to kill and sacrifice them before they are born, preventing them from suffering? On the other hand, is it possible to occupy their place? Is it possible to take the place of another human being? In other words, is it possible to unravel the thorn tangle in its place? What can the coach do if the tangle is too complex for him too? Is there anything too complex for a human being? What is more complex than its brain out there? How does one tame with love and then avoid any dilemmas? Can't human beings be like the hypothetical dogs that thirty thousand years ago began to get progressively closer, gaining much more confidence, until they started to live in marvellous mansions today? So, in this way until one enters the house of God? But, wouldn't a tamed human being be a pet of God? A dog that guards the garden by barking and attacking anyone who enters without permission? Is letting oneself be tamed to lose one's dignity? Then, is it right to face and question God, as Job did but to the last consequences? However, more significantly, does God want to tame or domesticate anybody? Just a few crackpots, one or two cranks? Does God want to make human beings something they are not, mutating or disguising them, a sort of plastic surgery? Otherwise, making them what they really are? Does God still want to create humanity, to raise them from scratch, from dust? Is the human being still nothingness, a zero to the left, only infinite potentiality, a bundle of illusions? Or rather, is God just seeking to turn human beings into nothing, making room to let them receive or perceive something divine in their hearts? Is it only God that can do this? In any case, what prevents God from doing so?

And so, in the last breath, what is the origin of the question? Is it the origin of life too? What is the origin of the answer? Is it the origin of death too? Is the question the origin from where a hidden answer emerges or is the answer what originates the question in order to hide itself? Anyway, is the answer always the expression of a programme, the word or the action of an automaton? A mechanical answer? Likewise, is the question always made by a machine too, a mechanical question? Is a *bioquestion* ever possible? That is, does a slave always ask, a reflection or a complex fold of a past structure always rooted in his or her own memory, a fig from a fig tree? If this is so, what is the daily life of the questioner and of the one who answers like, both tamed and domesticated by culture, science, tradition, knowledge, thought? By authorities, yesterday's wisdoms, desires or mysterious instincts? By most people's opinion, by a society like bees, a language, a natural or artificial selection that is pruning and orienting with or without sane objectives? Also by ignorance, illusions, stupidity, embodied lies, deceptions, speculations, and so on? Tamed and domesticated by other human beings, by fathers and mothers, brothers and sisters, wives and husbands, friends and strangers, politicians and priests, teachers and professionals, ideologies and philosophies, nations and countries, religions and moralities, genes and remembrances, natural laws and circumstances of life? By time and the environment, by a universe and its absolute or changing conditions, by a multiverse of selfish purposes and choices, by a mechanical or quantum mechanical God, by X, by a worldwide hypnotist, by death, by this and that, by an I, by you, by you yourself, by any possible pronouns, by one that conditions and enslaves oneself whatever one does and whatever one says, by a free and living God, by any God, by God Himself? Eh? ?

www.ingramcontent.com/pod-product-compliance
Lightning Source LLC
Chambersburg PA
CBHW051505170526
45166CB00001B/401